Victoria Guillén-Nieto
Hate Speech

Foundations in Language and Law

Volume 2

Victoria Guillén-Nieto

Hate Speech

Linguistic Perspectives

DE GRUYTER
MOUTON

Free access to the e-book version of this publication was made possible by the 16 institutions that supported the open access transformation *Purchase to Open Pilot* in collaboration with Jisc.

ISBN 978-3-11-161963-7
e-ISBN (PDF) 978-3-11-067261-9
e-ISBN (EPUB) 978-3-11-067276-3
ISSN 2627-3950
DOI https://doi.org/10.1515/9783110672619

Library of Congress Control Number: 2022947864

Bibliographic information published by the Deutsche Nationalbibliothek
The Deutsche Nationalbibliothek lists this publication in the Deutsche Nationalbibliografie; detailed bibliographic data are available on the internet at http://dnb.dnb.de.

This volume is text- and page-identical with the hardback published in 2023.
The book is published open access at www.degruyter.com

Cover image: kokouu/E+/Getty Images
Typesetting: Integra Software Services Pvt. Ltd.

www.degruyter.com

To Rebeca, my daughter, my love.

Acknowledgements

I could not have undertaken this journey without the guidance, support and encouragement of Janet Giltrow and Dieter Stein, editors of the Foundations in Law and Language series. Special thanks to Jurate Ruzaite, who read and commented in detail on this volume. I am grateful to Donato Mancini for his advice, suggestions and thorough copy-editing work. I would like to thank Fabienne Baider, Eckhard Bick and Matthias J. Becker for sharing their hate speech publications with me. Lastly, I thank Antonio Doval, Francisco Yus and Rebeca Ferrero for pointing me to the latest references in the field.

https://doi.org/10.1515/9783110672619-202

Preface

The linguistics of hate speech

Matsuda (1989) was the first legal scholar to use the term *hate speech*. Although Matsuda's initial examples only referred to *racial hate speech*, she clarified that, in her judgement, racial hate speech is not the only form of hate speech and that *anti-gay* and *anti-lesbian hate speech* are similar phenomena but deserve independent attention.

The meaning of the term hate speech is rather opaque, although, at first sight, it gives the opposite impression. Looking at the semantics of its constituent parts – that is, hate and speech – one may think that the term describes a subcategory of speech associated with the expression of hate or hatred towards people in general. However, we know by experience that its use is neither limited to speech nor to the expression of hatred. We also know that the target is not the general public but, instead, members of groups or classes of people identifiable by *legally-protected characteristics*, such as *race*, *ethnicity*, *religion*, *sex*, *sexual orientation*, *disability*,[1] amongst others. Therefore, the meaning of hate speech is not a function of the literal meanings of its constituent parts. On the contrary, it has multiple meanings, as suggested by Brown (2017a; 2017b).

The social phenomenon and the legal concept of hate speech are necessarily intertwined. As a social phenomenon, hate speech fuels broad-scale conflicts that may cause a breach of the peace or even create environments conducive to hate crimes. As a legal concept, hate speech is an abstract endangerment statute because it punishes not actual but hypothetical creation of social risk, and must find a balance between the fundamental rights to freedom of opinion and expression and dignity.

The analysis of hate speech is well documented in law, sociology and media and communication studies. For instance, in the field of law, Brown (2017a; 2017b) analysed the multiple definitions of hate speech; Tsesis (2002) examined how hate speech paves the way for harmful social movements and highlighted the destructive power of hate propaganda; Delgado (1982), Matsuda (1989) and Delgado and Stefancic (2018[2004]; 2018) claimed that the right to freedom of opinion and expression should be constrained by United States constitutional law when individuals abuse this right to shame, cause despair, inflict injury, threaten and harm members of groups or classes of people identifiable by legally-protected characteristics. Actionable hate speech within the legal framework of Spanish

1 Hall (2019) is one of the few studies on *disability hate speech*.

https://doi.org/10.1515/9783110672619-203

civil law was analysed by Dolz Lago (2015) and Landa-Gorostiza (2018). Disinformation and hate speech were studied, from an European Union constitutional law perspective, by Pitruzzella and Pollicino (2020). And Brown and Sinclair (2020) analysed the complex relationship between politics and hate speech laws.

Over the last few years, online hate speech has gained increasing interest amongst scholars in the fields of sociology and communication and media studies, as shown in the latest publications on the topic: *Online othering: Exploring digital violence and discrimination on the web* (Lumsden & Harmer 2019); *Digital hate: The global conjuncture of extreme speech* (Udupa, Gagliardone & Hervik 2021); *Social media and hate* (Banaji & Bhat 2022); and *Cyberhate in the context of migrations* (Monnier, Boursier & Seoana 2022).

On reviewing the literature, we find that while most of the research on hate speech principally comes from the social sciences, hate speech has not yet received sufficient attention as a scientific object of study in linguistics.[2] Only over the last few years can one see a marked turn in linguistics, especially in pragmatics, towards the analysis of language crimes, with special emphasis on (online) hate speech. The following references show that the linguistic analysis of hate speech is currently experiencing a boom: *A corpus linguistic analysis of white supremacist language* (Brindle 2016); *Language and violence: Pragmatic perspectives* (Silva 2017: 141–168); *Online hate speech in the EU: A discourse-analytic perspective* (Assimakopoulos, Baider & Millar 2017). This latter study, by Assimakopoulos, Baider and Millar, presents the findings of C.O.N.T.A.C.T., a project (2015–2017) co-funded by the Rights, Equality and Citizenship Programme of the European Commission Directorate-General for Justice and Consumers (JUST/2014/RRAC/AG) and coordinated by Baider at the University of Cyprus. Their project investigates online hate speech in the context of migration in the European Union from a critical discourse analysis perspective. The special issue of the journal *Lodz Papers in Pragmatics* (2018), edited by Baider and Kopytowska, focused on both hate speech and countering hate speech. In this issue, hate speech was analysed from different linguistic perspectives, such as pragmatics (Technau 2018: 25–43) and discourse analysis (Strani & Szczepaniak-Kozak 2018: 163–179), and approaches, especially corpus linguistics (Ruzaite 2018: 93–116). Millar (2019: 144–163) explored online hate speech and its social control from institutional and corporate perspectives.

2 In previous work, the present author made a similar claim regarding the analysis of other language crimes, such as defamation (Guillén-Nieto 2020), sexual harassment (Guillén-Nieto 2021) and workplace harassment (Guillén-Nieto 2022).

Linguists have also taken a step forward in providing technical services to the security forces and agents of internet platforms to detect and prevent online hate speech. In such cases, the methodology combines a corpus-based approach, deep learning and the qualitative linguistic analysis of lexical and grammatical indicators of hate speech. Some remarkable contributions to this cutting-edge research are represented by Becker (2020; 2021), who investigates antisemitism and the challenges that implicitness raises for automatic online hate speech detection; Bick (2020; 2021), who conducts computer-based research on the annotation of non-direct forms of hate speech, such as emoticons and emojis, in a German-Danish social media corpus;[3] and Baider (2020; 2022), who takes a legal-linguistic perspective on covert hate speech, combining The Rabat Plan of Action criteria with pragmatic analysis. Her work also draws attention to strategies for countering hate speech other than censoring it, as an efficient way to combat covert hate speech.

Lastly, *The grammar of hate: Morphosyntactic features of hateful, aggressive, and dehumanising discourse*, edited by Knoblock (2022), departs from traditional lexical and discursive approaches to hate speech detection. It focuses, instead, on the morphosyntactic features that the hate-advocating speaker appropriates, manipulates and exploits to express hate, hostility or violence toward the targets. The collection of chapters in this volume demonstrates how hate speech manifests itself in a wide array of grammatical features, such as morphology (Mattielo 2022: 34–58; Tarasova & Sánchez Fajardo 2022: 59–81), word formation (Beliaeva 2022: 177–196), gender (Lind & Nübling 2022: 118–139; Thál & Elmerot 2022: 97–117), pronouns (Flores Ohlson 2022: 161–176; Peterson 2022: 262–287), imperative verbs (Bianchi 2022: 222–240) and syntactic patterns (Geyer, Bick & Kleene 2022: 241–261), to name a few examples.

Why has hate speech not been studied in linguistics until recently? Linguistic theories have tended to address language as cooperative action (Grice 1975) geared to reciprocally informative polite understanding (Lakoff 1973; Brown & Levinson 1987 [1978]; Leech 1983). As a result of this idealised view of language, negative types of discourse and speech acts, such as defamation (Tiersma 1987; Shuy 2010; Guillén-Nieto 2020), harassment (Guillén-Nieto 2021; 2022; Stein 2022) and hate speech, have been traditionally cast aside as objects of linguistic study. Some linguists, such as Leech (1983), referred to impolite language as

3 The corpus has been developed within the framework of XPEROHS Project (2019–). *Towards a balance and boundaries in public discourse: Expressing and perceiving online hate speech.* Project leader: Klaus Geyer. https://xperohs.sdu.dk/publications/ (accessed 31 July 2022).

unusual, anomalous or deviant when it is, in effect, widespread. Since the 1990s, Culpeper (1996; 2005; 2008; 2011; 2012), Culpeper and Terkourafi (2016) and Kaul de Marlangeon (1993; 2005; 2008; 2014), amongst other linguists, have paved the way for the analysis of the deliberate use of language to offend. At present, the analysis of offensive communication is still making a place for itself within linguistic research. I concur with Knoblock (2022: 5) on the importance of giving scholarly attention to and discussing real world examples of hate speech openly, however offensive they might be, to improve understanding of how it functions and to seek feasible solutions.

This book is not a volume on hate speech laws nor an introduction to linguistics.[4] Instead, it is a volume on hate speech from various linguistic perspectives. In this sense, it is applied linguistics. As a linguistic object of study, hate speech is complex and, to a certain extent, elusive because it is not a unitary phenomenon (Brown 2017a; 2017b). Hate speech does not have a unified purpose. Hate speech can take permanent forms – e.g. racial epithets, insults, dehumanising metaphors, group defamation and negative stereotypes – but can also take transient forms. Hate speech can exist in various forms: written words, spoken words and audio-visual materials – e.g. gestures, symbols, images, films and video-games. Hate speech is not ascribed to any specific genre or rhetorical style, as it can range from thoughtful comments in a parliamentary speech to improvised sarcastic comments in an online post. Hate speech can involve many negative illocutionary and perlocutionary acts, such as insulting, degrading, humiliating, harassing, threatening, provoking, inciting hatred, hostility or violence and denying, justifying or glorifying acts of genocide. Hate speech is sometimes overt and unconcealed, but ever-increasingly coded and veiled (Becker 2020: 38). Hate speech can be delivered by identified speakers or can be anonymous, especially in online hate speech.

The book is divided into two different, but at the same time complementary, parts. Each part is devoted to one of the two applied linguistic disciplines in language and law or in law: *Legal linguistics* (Part I) and *Forensic linguistics* (Part II).[5] Legal linguistics analyses the doctrinal content of the law and its linguistically-based structure, while Forensic linguistics is concerned with helping to establish the facts on which a legal decision is based:

4 For an introduction to Linguistic pragmatics, see Alba-Juez and MacKenzie (2016).

5 For the differentiation between Legal linguistics and Forensic linguistics, see Guillén-Nieto and Stein (2022: 2–7).

> Forensic linguistics is the use of evidence from language use based on records, texts, or traces – not as the live substance, but as vestiges of the use of language, communication or speech acts that took place in the past, however medially constituted, spoken, written, digital, in connection with the resolution of crime (Guillén-Nieto & Stein 2022: 5).

Part I, *Legal linguistics*, consists of three chapters:

In Chapter 1. *Approaches to the meaning of hate speech*, the present author aims to demonstrate how Wittgenstein's concept of *family resemblance* can help our work on the description of the various definitions of hate speech. The discussion is divided into three parts, each corresponding to a different approach to hate speech. The first approach is ordinary language analysis, a philosophical investigation method concerned with how verbal expressions, in our case the term hate speech, is used in non-technical, everyday language (Wittgenstein 2009 [1953]). The discussion then focuses on various legal scholarly attempts to define hate speech: *content-based*, *intent-based* and *harms-based*. Finally, the chapter offers the reader a panoramic view of the existing regulations of hate speech in international law, common law and civil law – European Union law and Member States law.

In Chapter 2. *Hate speech as a legal problem*, the author deals with four significant issues at the core of hate speech as a legal problem: (a) the uneasy balance between the right to freedom of opinion and expression and the prohibition of incitement to hatred, hostility or violence, (b) the lack of an agreed-upon technical legal definition, (c) the difficulty of determining which speech acts are surface linguistic expressions of hate speech and (d) the legal challenges raised by the advent of online hate speech.

Chapter 3. *The legal reasoning in hate speech court proceedings* is devoted to the key foci of legal reasoning about hate speech across different jurisdictions, specifically the United States and the European Union. The chapter begins with a review of several landmark decisions of the United States Supreme Court and the United States Court of Appeals for the Armed Forces. In parallel, some landmark decisions of the European Court of Human Rights are also reviewed. Finally, the chapter analyses what constitutes evidence of hate speech in the United States and the European Union, which have been taken as representative examples of legal practices in common law and civil law jurisdictions.

Part II, *Forensic linguistics*, targets the language cues that various linguistic theories can provide for making hate speech legally actionable. The discussion aims to demonstrate the value of a micro-language approach to hate speech. In order not to build a castle in the sky, the analysis is grounded in several relevant legal cases selected from those presented in Part I (Chapter 3).

Part II consists of five chapters:

Chapter 4. *Critical discourse analysis* reviews some central theories in CDA that are deemed useful for an improved understanding of hate speech: (a) the *theory of social representations* (van Dijk 1997; 2005; 2006b), (b) the *theory of ideology* (van Dijk 1995a; 1995b) and (c) the *theory of power as control* (van Dijk 1996; 2015). The chapter illustrates the benefits of a multilevel analysis (macro level, meso level and micro level) through its application to a case associated with racial hate speech: Brandenburg v. Ohio (1969). I argue that a CDA approach may help unveil the social and discourse practices reproducing racism in Brandenburg v. Ohio (1969) in the eyes of the law.

In Chapter 5. *Register and genre perspectives on hate speech*, the author elaborates on the surface linguistic forms articulating hateful texts. Specifically, the chapter focuses on the texts – discourse segments of various dimensions – in which hate speech manifests itself and on the genre or genres into which such texts can be classified. My purpose is to demonstrate to what extent the register and genre perspectives can improve our understanding of hate speech.

Chapter 6. *Speech act theory* investigates the types of speech acts that give expression to hate speech. Although the application of Speech act theory to the analysis of hate speech is problematic, due to the inherent complexity of hate speech as an empirical object of study, the chapter points to the useful insights the theory can still provide at both macro and micro levels of linguistic analysis. The author draws the reader's attention to implicitness, indirectness and a loose illocutionary-perlocutionary link as some of the major problems in recognising and identifying the speech acts giving expression to hate speech.

Chapter 7. *(Im)politeness theory* points to the insights this socio-pragmatic theory can provide in the analysis of hate speech. The author argues that to understand the hate-advocating speaker's impolite linguistic behaviour (Culpeper 2011), one needs, first, to analyse their intentional deviation from polite behaviour. The chapter reviews the main approaches to politeness: (a) the *conversational-maxim approach* (Lakoff 1973; Leech 1983) and (b) the *face-saving approach* (Brown & Levinson 1987 [1978]). The discussion then moves forward to *impoliteness* (Culpeper 2011; Kaul de Marlangeon 2005). The chapter also illustrates how the various theoretical views on (im)politeness can be applied to actual legal cases associated with hate speech.

In Chapter 8. *Cognitive pragmatics*, the author deals with the meaning and interpretation of hate speech. Specifically, the discussion concentrates on how the hate-advocating speaker communicates an intention that is not explicitly stated and how this intention is likely to be interpreted differently by subjects not belonging to the same interpretive community. Grice's (1975) *conversational implicature* provides the bridge from what is said to what is meant but not overtly said. The interpretation of the speaker's intended meaning is considered through Sperber and Wilson's (1995 [1986]) *Relevance theory*, with special reference to *ostensive-inferential communication*, with the aim of demonstrating the extent to which cognitive pragmatics can help unveil the hate-advocating speaker's intended meaning.

As mentioned at the outset of this Preface, hate speech is a complex object of linguistic analysis. Since hate speech is not a unitary phenomenon but, instead, multi-layered, and indeed multi-modal, it cannot be wholly explained from a single linguistic perspective. On the contrary, hate speech demands successive analyses, each focusing on a specific linguistic element. For this reason, the reader will see that in the chapters of Part II, the author sometimes recurs to certain landmark legal cases associated with hate speech, especially Terminiello v. Chicago (1949) and Brandenburg v. Ohio (1969), to illustrate the insights that each linguistic theory may provide into the same case. In this respect, hate speech resembles the elephant in the well-known poem *The blind men and the elephant:*[6] A group of six blind men went to see an elephant. They thought that by observation, they could learn the elephant's appearance. Each blind man felt a different part of the elephant's body: the side, the tusk, the trunk, the knee, the ear and the swinging tail. As a result, there were six different descriptions of the elephant. Depending on the part felt by each blind man, elephant was like a wall, a spear, a snake, a tree, a fan or a rope. The blind men's descriptions of the elephant were different from each other because they were based on their own single experiences. All of them were partly right but they were all wrong. From this fable one can learn that if one takes a single perspective to look at hate speech – our elephant – one can only describe one of its linguistic elements, ignoring other elements that may be equally relevant to establish the facts in the judicial narrative. In Part II, the reader will be able to discover, through successive linguistic analyses, several surface linguistic manifestations of hate speech, especially the register, the genre, the speech acts, the strategies and the ostensive stimuli employed by hate-advocating speakers to signal their malicious communicative intents.

---

6 The American poet John Godfrey Saxe (1816–1887) is believed to have introduced a Hindu fable to western readers with the poem *The blind men and the elephant* (cf. Saxe, John Godfrey. 1876. *The Poems of John Godfrey Saxe*, 259–261. Boston). The "blind men" in the poem do not represent real blind people. Nor does "the elephant" in the poem represent a real elephant. On the contrary, "the blind men" and "the elephant" are the fictional characters of a fable whose moral is that humans tend to claim absolute truth based on their limited, subjective experience as they ignore other people's limited, subjective experiences, which may be equally true.

This book also evidences the symbiosis between Linguistics and Forensic linguistics in which both sides benefit from the relationship. Linguistics provides Forensic linguistics with theories, methodologies and tools for analysing hate speech. In contrast, Forensic linguistics prompts adjustments and advances in linguistic pragmatic theories and methodologies because of the new complex input data provided by language crimes, in this case, hate speech.

Contents

1 Approaches to the meaning of hate speech

1 Introduction

The term *hate speech* recognises a wide range of harmful social practices and discourses. Let us take, for instance, Islamophobic blogs, cross burnings, racial epithets, or dehumanising pictures of Jews. If all these social activities fall under hate speech, they must have certain elements or features in common. It might therefore be reasonable to think that a definition of hate speech should comprise the essential elements that enable us to recognise the phenomenon. This chapter aims to demonstrate how Wittgenstein's concept of *family resemblance*, which Brown[7] borrowed in his legal essay *What is hate speech? Part II: Family resemblances* (2017b), can help linguists and legal scholars better describe hate speech. The discussion is divided into three parts, each corresponding to a different approach to hate speech. The first approach I deal with is *ordinary language analysis*, a philosophical investigation method concerned with how verbal expressions, in our case, the term hate speech, is used in non-technical, everyday language (Wittgenstein 2009 [1953]). Then I analyse various legal scholarly attempts to define hate speech. Finally, the chapter provides an overview of the approaches to a technical legal definition of hate speech in international law, common law and civil law.

2 Wittgenstein's concept of *family resemblance* applied to the definition of hate speech

Family resemblance is a philosophical concept found in Wittgenstein's posthumously published *Philosophical investigations* (2009 [1953]). Wittgenstein turned to games, as a way to explain the gist of his idea:

> Look, for example, at board games, with their various affinities. Now pass to card games; here, you find many correspondences with the first group, but many common features drop out, and others appear. When we pass next to ball games, much that is common is retained, but much is lost [. . .] And the upshot of these considerations is that we see a complicated network of similarities overlapping and crisscrossing; similarities in the large and in the small (Wittgenstein 2009 [1953]: 36).

7 Alexander Brown is a Reader in Political and Legal Theory at the University of East Anglia in the United Kingdom.

https://doi.org/10.1515/9783110672619-001

The application of the concept of family resemblance to the various definitions of hate speech involves, on the one hand, comparing disparate definitions and identifying sets of overlapping and crisscrossing similarities between those definitions. The essential idea is that any definition of hate speech shares at least one similar element with at least one other definition, even if there is no single element common to all definitions. As a result, as Brown argued, "[. . .] we are likely to end up with a family of different meanings of the term hate speech, each slightly different from the next but sharing at least one essential element in common" (Brown 2017b: 13).

3 Ordinary language analysis: An approach to understanding the meaning of hate speech in everyday language

Wittgenstein (2009 [1953]) laid the groundwork for ordinary language analysis: he advocated that, in solving philosophical problems, understanding how language is used in context is more important than its abstract meaning. In "What is hate speech? Part 2: Family resemblances", Brown (2017b) applies ordinary language analysis to describe how the term hate speech is used in non-technical, everyday language. He divides his analysis into four parts: (1) *purpose-oriented analysis*, (2) *folk platitudes analysis*, (3) *intuitions analysis* and (4) *everyday use analysis*. In the following, I summarise the major insights Brown provides in each type of analysis.

1) Purpose-oriented analysis. The term hate speech may be used for five different purposes in non-technical, everyday language: (a) highlighting forms of harmful speech, (b) pointing out socially disruptive forms of speech, (c) identifying forms of speech that can challenge people's sense of equality, (d) articulating norms of civility and (e) labelling forms of speech that undermine democracy.
2) Folk platitudes analysis. There seem to be four main commonplace generalisations, which Brown (2017b) calls folk platitudes, held by people about the meaning of the term hate speech: (a) it is negatively evaluative; (b) it is directed against groups of people identifiable by legally-protected characteristics – e.g. race, ethnicity, religion, sexual orientation, gender identity and disability; (c) it relates to emotions, feelings, or attitudes that can lead to incitement to hatred, and are even liable to trigger violence; and d) it is speech that is not protected by an individual's right to freedom of opinion and expression.
3) Intuitions analysis. People seem to have certain intuitions about the forms of speech that may be associated with hate speech. Most people would agree

on classifying the following forms of speech as hate speech when they are addressed to historically oppressed or victimised groups: (a) insults, racial epithets and ethnic jokes (Leader, Muller & Rice 2009); (b) any forms of speech that articulate ideas relating to the moral inferiority or non-humanity; (c) presenting false statements that harm social reputation; and (d) words or behaviour that threaten, support, or incite hatred or violence, or words that simply justify or glorify violence.

4) Everyday language. There seem to be several speech acts and discourses that give expression to hate speech, such as racial hate speech and homophobic hate speech.

In sum, Brown's (2017b) ordinary language analysis approach to the meaning of the term hate speech demonstrates that in non-technical, everyday language, the term does not have a single meaning but, instead, a family of meanings, each slightly different from another but sharing at least one essential element in common.

4 Legal scholarly attempts to defining the concept of hate speech

Some significant legal scholarly attempts to define hate speech began in the 1980s and continue today. Marwick and Ross argued that one can broadly define hate speech as "[. . .] speech that carries no meaning other than hatred towards a particular minority, typically a historically disadvantaged minority" (Marwick & Ross 2014: 17). Marwick and Ross identified three distinct categories that legal scholars use to define hate speech:

1) Content-based hate speech includes words, expressions, symbols and iconographies, such as the Nazi swastika or the Ku Klux Klan's burning cross, generally considered offensive to a particular group of people and objectively offensive to society.
2) Intent-based hate speech requires the speaker's communicative intention to incite hatred or violence against a particular minority, member of a minority, or person associated with a minority without communicating any legitimate message.
3) Harms-based hate speech is speech that causes the victim harm, such as loss of self-esteem, physical and mental stress, social and economic subordination and effective exclusion from mainstream society.

In the following sections, I will review some of the most significant definitions in the legal scholarly domain and classify them according to the above categories or elements identified by Marwick and Ross (2014).

4.1 Content-based hate speech

In the late 1980s, a group of legal scholars in the United States used the term "racist speech" in response to how different legal systems tackled a particular type of harmful speech targeted against specific racial groups. In his influential work, *Words that wound: A tort action for racial insults, epithets, and name-calling* (1982), Delgado opened the path to scholarly discussion about why racist speech should be considered a civil tort under US law. Delgado (1982: 179) proposed an element-based definition that requires racist speech to be actionable if: (1) the language that was addressed to the plaintiff by the defendant intended to demean through reference to race, (2) the plaintiff understood the language as intended to demean through reference to race and (3) a reasonable person would recognise the language employed by the defendant as a racial insult. Delgado, then, defined racist speech as a specific type of civil tort. Accordingly, his definition comprised the three essential elements in a civil tort, with the only difference being that the offence revolves around insult on racial grounds: (1) an intent-based element to cause harm through racial insult, (2) the plaintiff must secure uptake of the intended racial insult and (3) the act must be, objectively, a racial insult.

In the same vein, Matsuda (1989) claimed that US law should provide legal means for bringing action against hate-advocating speakers in America by criminalising racist speech because of its harmful effects on the targeted racial minorities and on democratic social values. Matsuda's standpoint had as its base the provisions of international law such as the Universal Declaration of Human Rights,[8] the International Covenant on Civil and Political Rights (ICCPR)[9] and the International Convention on the Elimination of All Forms of Racial

8 The United Nations General Assembly proclaimed the Universal Declaration of Human Rights as a common standard of achievements for all peoples and all nations in Paris on 10 December 1948 (General Assembly resolution 217 A). https://www.un.org/en/universal-declaration-human-rights/ (accessed 31 May 2020).

9 The International Covenant on Civil and Political Rights. Adopted by the General Assembly of the United Nations on 19 December 1966. https://treaties.un.org/doc/publication/unts/volume%20999/volume-999-i-14668-english.pdf (accessed 26 May 2020).

Discrimination (CERD).[10] It should be noted that Matsuda did not make any distinctions between racist speech, hate speech, hateful speech and hateful messages. She used the terms interchangeably to refer to the same social phenomenon. Although, in Matsuda's work, the term hate speech primarily refers to racism, she also recognised other types of hate speech based on sex and sexual orientation, such as anti-gay and anti-lesbian hate speech, that, in her judgement, require independent attention. For Matsuda, racial hate speech and hate speech on the grounds of sexual orientation are products of different types of discourses. Specifically, whereas racial hate speech is a product of a discourse of subordination that claims the superiority of the white race over the black race, hate speech on the grounds of sex and sexual orientation is a product of a discourse of power that claims the supremacy of masculinity and heterosexuality over femininity and homosexuality. Matsuda (1989: 2358) suggested that there are three elements at the core of hate speech: (1) the message is of racial inferiority, (2) the message is directed against a historically oppressed racial group and (3) the message is hateful and degrading. Unlike the definition of Delgado, Matsuda's seems to be more effect-oriented in the first and third elements of the definition – that is, "the message is of racial inferiority", "the message is hateful and degrading". Matsuda's definition also highlights that hate speech must be "directed against historically oppressed groups".

In her work, Matsuda considered the violence of the word in *threats*, *racial epithets*[11] and *slurs*,[12] and the violence of *symbols*, such as the Klan burning crosses and their robes, which take their hateful meaning from their historical context and connection to violence. Matsuda's approach foregrounded the damaging effects hate speech may have on social peace because it is a speech that causes hatred, persecution and degradation of certain groups.

Parekh (2006) drew attention to the difficulty of recognising hate speech because of its close and complex association with conflict[13] (Nelde 1987:

---

10 The International Convention on the Elimination of All Forms of Racial Discrimination. Adopted and opened for signature and ratification by General Assembly resolution 2106 (XX) of 21 December 1965, entry into force 4 January 1969, under Article 19. https://www.ohchr.org/en/professionalinterest/pages/cerd.aspx (accessed 3 June 2020).

11 A racial epithet, also known as a racial slur, is a derogatory term or expression based on someone's racial background. Racial epithets are understood to convey contempt and hatred toward their targets (cf. Hom 2008).

12 A slur is an insulting remark that could damage someone's reputation.

13 Language conflict involves intrastate political tension or civil unrest between speakers of different lingua-cultures. Language conflict plays a central role in group and national identity; hence, it is a surface indicator of entrenched political, social and economic conflict.

33–42; Darquennes 2015: 7–32; Millar 2019: 145–163), offence, abuse and discrimination (Stollznow 2017). Hate speech can be triggered by social conflict, but hate speech itself can also spark conflict, violence and a breach of the peace. Hate speech can be conveyed through offensive,[14] abusive[15] and other types of harmful speech. It must be clarified that these types of speech, which should be considered broad categories including many types of rude, nasty, malicious or even humiliating expressions, are only manifestations of hate speech under specific circumstances – e.g. the hateful message must be addressed to a legally-protected group.

Parekh argued that the reasons for prohibiting hate speech are specific and should be distinguished from those restricting other types of language crimes, such as defamation (libel and slander) and threats. Parekh (2006: 214) proposed a definition of hate speech based on three essential elements: (1) it singles out an individual or a group based on legally-protected characteristics, (2) it stigmatises the target group and (3) it marginalises the target group. Parekh's definition focused on the discriminatory content hate messages convey, as it stigmatises, dehumanises and demonises the members of the target group with the intent of placing them outside of public life. For Parekh, it is the hateful content of the messages, rather than the immediate consequences, that should matter to the law. In Parekh's words:

> It is, therefore, a mistake commonly made to define hate speech as one likely to lead to public disorder and to proscribe it because or only when it is likely to do so. What matters is its content, what it says about an individual or a group, not its likely immediate consequences, and our reasons for banning it need not be tied to the latter (Parekh 2006: 214).

Since Parekh's definition does not provide an intent-based element, one can expect it to attract severe criticism from freedom of speech theorists. Apart from pointing to the harmful content of hate speech, Parekh (2006: 214) also laid stress on the multimodal nature and pragmatic meaning of hate speech forms. Parekh argued that although speakers often convey hatred through explicit, offensive and abusive language, hate speech can also be expressed in implicit and moderate language. For instance, hate speech can be embedded in maliciously ambiguous jokes, implicatures and non-verbal elements of communication (see Chapter 8). From Parekh's insights, one can learn that although a violent mode of expression provides a useful clue to the malevolent nature of a

---

14 Offensive language refers to hurtful, derogatory or obscene comments by one person to another.

15 Abusive language refers to harsh, violent, profane, or derogatory language that could demean the dignity of an individual.

message, it is inadequate to take overtly violent messaging as the only clue. One should also be cautious in associating hate speech with abusive language, because this type of negative language can also be used to describe other language crimes, such as insults, defamation (libel or slander) and threats. For Parekh (2006), the meaning of hate speech can only be understood in the context of situation. An utterance may appear to be innocent in one context, while the same utterance may not be so innocent in another context. This ambiguous condition would be a hindrance to the elaboration of a closed catalogue or inventory of hate speech forms.

Waldron (2012) also proposed a content-based definition of hate speech. In his view, hate speech refers to group defamation or group libel because the messages may involve several reputational attacks that amount to "assaults upon the dignity of the person affected" (Waldron 2012: 59). For Waldron, a message may qualify as hate speech if it involves "some imputation of terrible criminality" (Waldron 2012: 47) that can affect the human dignity of people who belong to a legally-protected group.

4.2 Intent-based hate speech

Moran (1994) defined hate speech as "speech that is intended to promote hatred against traditionally disadvantaged groups" (Moran 1994: 1430). Moran's definition introduced a new element: the perpetrator's malicious intent to "promote hatred", and it also broadened the scope of possible target groups because the wrong action is not exclusively directed against racial groups. Ward (1997), for his part, claimed that the legal restraint of hate speech is a means to strengthen the individual's right to freedom of opinion and expression. He defined hate speech as "any form of expression through which speakers intend to vilify, humiliate, or incite hatred against their targets" (Ward 1997: 765). His definition includes three elements: (1) the subject's intent to damage the target's rights to human dignity and equality, (2) the subject's intent to incite hatred, which involves urging or persuading someone to act violently or unlawfully against the target and (3) the message must be perceived as a *verbal attack* against the target groups.

In the same line of thought, Benesch described a specific subcategory of hate speech that she named *dangerous speech* (Benesch 2014: 22) because of its power to call for violence against the target groups. Benesch's dangerous speech resembles Matsuda's *fighting words* (1989: 2355) and Ward's *verbal attack* (1997: 766) in that the term foregrounds the element of intent to incite hatred or violence against the targets.

4.3 Harms-based hate speech

Drawing on Matsuda's (1989) definition of racist speech, Massey (1992) provided a broad and detailed argument on the legal reasons for suppressing hate speech based on the foundational paradigms of free speech. Massey's definition of hate speech exclusively focuses on the harm it causes to freedom of opinion and expression. In his view, hate speech should be actionable if it (1) causes harm or violence to individuals belonging to specific groups and (2) abuses the right to freedom of opinion and expression by inflicting injury on and causing despair to its targets. In essence, Massey's definition shares with Matsuda's one main idea: the principles of racist ideology are irreconcilable with the foundational paradigms of equality and freedom of speech of the Constitution of the United States. For this reason, according to Massey, hate speech should be considered restricted speech and regulated by law.

After reviewing the three main legal scholarly approaches to the definition of hate speech, content-based, intent-based and harms-based, it is clear that hate speech is not a univocal concept that can be defined unambiguously, but rather a multi-layered concept with multiple meanings that requires various approaches. As a result, one has a family of legal scholarly definitions of hate speech, each slightly different from the other in perspective but sharing at least one core element with another definition. Therefore, I can also conclude that Wittgenstein's concept of family resemblance (2009 [1953]) improves our understanding of the various legal scholarly approaches to the definition of hate speech.

5 Approaches to a technical legal definition of hate speech

Since World War II, many nations have passed laws to curb incitement to hatred against minorities. These laws, which started as protections against anti-Semitism, are known under the rubric of hate speech laws. The elements of hate speech, as in the case of specific civil or criminal offences, must be captured in their legal definition, which enables prosecutors, judges and juries to examine the offence in light of the specific circumstances of a given legal case. However, the problem, as the next sections show, is that there is no agreed-upon technical legal definition of hate speech. The discussion that follows then aims to provide good examples of the differing approaches to a technical legal definition of hate speech at three levels: international law, common law and civil law – European Union law and Member State law.

5.1 International law

The Universal Declaration of Human Rights (UDHR) (1948) is at the core of hate speech laws because it set out the principle, for the first time, that there must be universal protection for people's fundamental rights. Since hate speech involves vicious acts against human dignity and equality, it violates the fundamental rights articulated in Article 2:

> Everyone is entitled to all the rights and freedoms outlined in this Declaration, without distinction, such as race, colour, sex, language, religion, political opinion, national or social origin, property, birth or status. Furthermore, no distinction shall be made based on the political, jurisdictional or international status of the country or territory to which a person belongs, whether it be independent, trust, non-self-governing or under any other limitation of sovereignty.

In the same vein, the International Convention on the Elimination of All Forms of Racial Discrimination (CERD) (1965) imposes responsibility on states for the prohibition of hate speech by law in Article 4:

> States Parties condemn all propaganda and organisations disseminating ideas or theories of superiority of one race or group of persons of one colour or ethnic origin, or which attempt to justify or promote racial hatred and discrimination in any form and undertake to adopt immediate and positive measures designed to eradicate all incitement to, or acts of, such discrimination and, to this end, with due regard to the principles embodied in the Universal Declaration of Human Rights.

Hate speech must be prohibited by law because it infringes on universal fundamental rights, such as the rights to human dignity and equality. The International Covenant on Civil and Political Rights (ICCPR) (1966) refers to hate speech in Article 20, paragraph 2), as

> Any advocacy of national, racial or religious hatred that constitutes incitement to discrimination, hostility or violence shall be prohibited.

Hate speech, then, consists of three essential elements that must co-occur:
1) Advocacy: Public forms of expression intended to elicit action or response.
2) Hatred: Intense negative emotions towards a group of people identified by legally-protected characteristics.
3) Incitement: Public forms of expression likely to trigger acts of discrimination, hostility or violence against a group of people identified by legally-protected characteristics.

Although there is no agreed-upon technical legal definition, both the International Convention on the Elimination of All Forms of Racial Discrimination

(CERD) (1965) and the International Covenant on Civil and Political Rights (ICCPR) (1966) share three elements in their definition of hate speech: (1) perpetrators use public forms of expression, (2) perpetrators target victims belonging to a historically disadvantaged group of people and (3) perpetrators have a malicious intent. Both the International Convention on the Elimination of All Forms of Racial Discrimination (CERD) (1965) and the International Covenant on Civil and Political Rights (ICCPR) (1966) identify the element of intent. However, their approaches to hate speech are different. Whereas the former refers to a wider array of speech acts, such as “disseminate”, “justify”, “promote” and “incite” racial hatred and discrimination against the targets, the latter unambiguously focuses on the intent “to trigger acts of discrimination, hostility or violence” and its scope is not limited to racial hatred.

5.2 Common law

5.2.1 Hate crime legislation in the United States

At the federal level, the Hate Crime Statistic Act does not refer to hate speech but instead to hate crimes in the following terms:

> Under the authority of section 534 of title 28, United States Code, the Attorney General shall acquire data, for each calendar year, about crimes that manifest evidence of prejudice based on race, gender and gender identity, religion, disability, sexual orientation, or ethnicity, including where appropriate the crimes of murder, nonnegligent manslaughter; forcible rape; aggravated assault, simple assault, intimidation; arson; and destruction, damage or vandalism of property.[16]

US law therefore prohibits hate crimes against individuals who belong to legally-protected groups. Unlike hate crime, hate speech does not have a legal definition under US law and enjoys substantial protection under the First Amendment to the Constitution of the United States. According to the First Amendment:

> Congress shall make no law respecting an establishment of religion, or prohibiting the free exercise thereof; or abridging the freedom of speech, or of the press, or the right of the people peaceably to assemble and to petition the government for a redress of grievances.[17]

16 Hate Crime Statistics Act, As Amended, 28 USC. § 534 § [Sec. 1.]. https://ucr.fbi.gov/hate-crime/2017/resource-pages/hate-crime-statistics-act.pdf (accessed 27 May 2020).
17 The Constitution of the United States. First Amendment. https://constitution.congress.gov/constitution/amendment-1/ (accessed 27 May 2020).

The First Amendment to the Constitution of the United States protects speech no matter how offensive its content may be. This protection entails from the assumption that the fundamental right to freedom of opinion and expression requires the government to strictly protect vigorous debate on matters of public concern even when such debate devolves into a dangerous speech that causes other people to feel grief, anger, or fear. This does not mean that the First Amendment does not provide any restrictions to hate speech. It does, in effect, impose limitations on forms of expression such as "genuine threats", "defamatory speech", "obscenity" and "incitement to imminent violence". An individual's speech may be restricted in the United States if (1) it is intended to produce lawless action, (2) it is likely to incite such action and (3) such action is likely to occur imminently. This set of restrictions to freedom of opinion and expression stem from the case of Brandenburg v. Ohio (1969), whose final judgment is known for having replaced the *clear and present danger standard* with the *imminence standard*. In this case, Brandenburg, a Ku Klux Klan leader, was convicted by the Court in Ohio for delivering a protest speech at a Klan rally, in which he advocated violence against African Americans and Jews. The Supreme Court did not only overturn Brandenburg's conviction but also struck down the Ohio statute prohibiting his speech because

> Freedoms of speech and press do not permit a State to forbid advocacy of the use of force or law violation except where such advocacy is directed to inciting or producing imminent lawless action and is likely to incite or produce such action (Brandenburg v. Ohio (1969). No. 492).[18]

As shown in the above excerpt from the Supreme Court's judgment in Brandenburg v. Ohio (1969), hate speech in US law strictly refers to that speech that "is directed to inciting or producing imminent lawless action, and it is likely to incite or produce such action." In practice, this definition imposes a very high standard for a case of hate speech to be declared in US courts of justice. The definition also confirms the protected status of the right to freedom of opinion and expression under the First Amendment to the Constitution of the United States.

5.2.2 Hate speech legislation in Canada

Under Canadian law, hate speech refers to speech that publicly and wilfully promotes hatred against groups of people that can be identified by legally-

18 US Supreme Court. Brandenburg v. Ohio, 395 US 4444 (1969). Brandenburg v. Ohio. No. 492. Argued February 27, 1969. Decided June 9, 1969. https://supreme.justia.com/cases/federal/us/395/444/ (accessed 18 August 2020).

protected characteristics, as stated in section 318(4) of the Criminal Code.[19] In addition, at a provincial level, most provinces contain provisions in their respective human rights legislation that prohibit forms of expression that incite hatred or violence against individuals or groups identified by legally-protected characteristics.

5.2.3 Hate speech legislation in the United Kingdom

Although there is no specific hate speech law in the United Kingdom, several laws forbid incitement to hatred or violence against legally-protected groups. The hate crime legislation in the United Kingdom criminalises conduct that is intended or likely to stir up racial hatred or is intended to do so based on religion or sexual orientation. The term "conduct" includes the use of hateful words and a wide range of forms of expression, such as displays of text, books, banners, photos and visual art, the public performance of plays and the distribution or presentation of pre-recorded material. The legal concept of "stirring up offences" is an effort to challenge racial hatred in the United Kingdom. Racial hatred, which was initially enacted in the Race Relations Act 1968,[20] is defined in the Public Order Act 1986 as "hatred against a group of persons defined by reference to colour, race, nationality (including citizenship) or ethnic or national origins."[21] Sections 18 to 22 of the Public Order Act 1986 set out several acts through which racial hatred is likely to be stirred up. For example, by using words, behaviour or displaying written material (s. 18), publishing or distributing material (s. 19), publicly performing a play (s. 20), distributing, showing or playing a recording (s. 21), broadcasting or including a programme in cable programme service (s. 22).

Over the last few years, new incitement offences have joined the Public Order Act 1986 – e.g. religious hatred joined in 2006 and incitement to hatred on sexual orientation grounds joined in 2008. The legislation defines religious hatred as "hatred against a group of persons defined by religious belief or lack of religious belief",[22] and hatred on the grounds of sexual orientation as "hatred against a group of persons defined by sexual orientation (whether towards persons of the

---

19 Criminal Code, R.S.C. 1985, c.C-46, s. 318(4). https://laws-lois.justice.gc.ca/eng/acts/c-46/page-68.html#docCont (accessed 27 May 2020).

20 Race Relations Act 1968 c.71. Part I. Discrimination. http://www.legislation.gov.uk/ukpga/1968/71/enacted (accessed 27 May 2020).

21 Public Order Act 1986. Part III. Racial Hatred, s.17. Meaning of racial hatred. http://www.legislation.gov.uk/ukpga/1986/64/contents (accessed 27 May 2020).

22 Public Order Act 1986. Part III. Racial Hatred, s. 29A. Meaning of religious hatred. http://www.legislation.gov.uk/ukpga/1986/64/contents (accessed 27 May 020).

same sex, the opposite sex or both sex)".[23] Sections 29B to 29F set out acts intended to stir up religious hatred or hatred on the grounds of sexual orientation. These acts are the same as those set out for racial hatred.

5.2.4 Hate speech legislation in Australia

In Australia, the Racial Discrimination Bill 1975 (RDB) declared the prohibition of offensive behaviour motivated by others' race, colour or national or ethnic origin. Specifically, section 18C states:

It is unlawful for a person to do an act, otherwise than in private, if:

1) the act is reasonably likely, in all the circumstances, to offend, insult, humiliate or intimidate another person or a group of people; and
2) the act is done because of race, colour or national or ethnic origin of the other person or some or all people in the group.[24]

Another relevant law in Australian legislation in which one can find an attempt to identify the punishable elements of hate speech is the Criminal Code Amendment (Racial Vilification Act), which is a hate speech law that uses the terms "racial animosity" and "racist harassment" rather than hate speech. The Racial Vilification Act describes unlawful conduct in sections 77–80D, ranging from conduct intended to incite hatred against targeted racial groups, dissemination of material for incitement to racial hatred and racial harassment. Specifically, section 77 defines conduct intended to incite racial animosity or racial harassment as

> Any person who engages in any conduct, otherwise than in private, by which the person intends to create, promote or increase animosity towards, or harassment of, a racial group or a person as a member of a racial group is guilty of a crime and is liable to imprisonment for 14 years.[25]

The above definition refers to three essential elements of racial hate speech: (1) it involves malicious conduct directed against a racial group, (2) it requires a specific intent to commit the crime and (3) the purpose of this type of harmful conduct is to incite racial animosity or racial harassment towards a racial group.

---

23 Public Order Act 1986. Part III. Racial Hatred, s. AB. Meaning of hatred on the grounds of sexual orientation. http://www.legislation.gov.uk/ukpga/1986/64/contents (accessed 27 May 2020).

24 The Racial Discrimination Bill 1975. https://www.legislation.gov.au/Details/C2014C00014 (accessed 28 May 2020).

25 The Racial Vilification Act 2004. Section 77. https://www.legislation.wa.gov.au/legislation/statutes.nsf/law_a9290_currencies.html (accessed 28 May 2020).

One significant difference between common law legal systems is the specific weight they give to the right to freedom of opinion and expression over the right to dignity and equality. On the one hand, the hate speech legislation in the United Kingdom and Australia seem to favour the fundamental rights to dignity and equality over the right to freedom of opinion and expression. UK law prohibits speech that stirs up hatred on the grounds of race, religion and sexual orientation, while Australian law prohibits racial animosity and racial harassment. In these two common law jurisdictions, hate speech is related to the incitement to hatred, especially racial hatred or racial harassment, and it is this type of restricted speech that the law prohibits without considering whether it may, or may not, lead to violent acts against the target group. On the other hand, hate speech laws in the United States and Canada seem to favour the right to freedom of opinion and expression over the right to dignity and equality when they only ban speech that is "likely to incite or produce imminent lawless action" (US law) or is "likely to cause a breach of the peace" (Canadian law).

5.3 Civil law

5.3.1 Hate speech in European Union law

Hate-speech legislation in Europe goes back to the European Convention on Human Rights.[26] Specifically, Article 14 prohibits discrimination based on sex, race, colour, language, religion, political opinion, national or social origin, association with a national minority, property, birth or status. The fundamental rights to dignity and equality prevail over the right to freedom of opinion and expression under European Union law. Recommendation No. R (97) 20 of the Committee of Ministers of the Council of Europe to the Member States, adopted on 30 October 1997, defines hate speech as

> all forms of expression which spread, incite, promote or justify racial hatred, xenophobia, anti-Semitism or other forms of hatred based on intolerance, including intolerance expressed by aggressive nationalism and ethnocentrism, discrimination and hostility against minorities, migrants and people of immigrant origin (Committee of Ministers of the Council of Europe 1997: 107).[27]

---

26 European Convention on Human Rights. Convention for the Protection on Human Rights and Fundamental Freedoms. Rome, 4. XI, 1950. https://www.echr.coe.int/Documents/Convention_ENG.pdf (accessed 27 May 2020).

27 Council of Europe Recommendation No. R (97) 20 of the Committee of Ministers to the Member States on Hate Speech (Adopted by the Committee of Ministers on 30 October 1997, at

The above definition presents two important differences compared with the common law. First, its scope is broader because, apart from "incitement to hatred", it considers other types of acts, such as "spread", "promote" and "justify". Second, the target groups are more varied, as they refer to racial hatred and xenophobia, anti-Semitism and other hate-based intolerance against minorities and migrants. In the same vein, the Council Framework Decision (2008),[28] on combating certain forms and expressions of racism and xenophobia through criminal law, defines hate speech as certain forms of conduct which are punishable as criminal offences, such as (1) public incitement to violence or hatred directed against a group of persons or a member of such a group on the grounds of race, colour, descent, religion or belief, or national or ethnic origin; (2) public dissemination or distribution of pictures or other material; (3) publicly condoning, denying or grossly trivialising crimes of genocide, crimes against humanity and war crimes; and (4) instigating, aiding or abetting in the commission of the offences mentioned in (1), (2) and (3).

The definition provided by the Council Framework Decision (2008) comprises a wider range of forms of expression, apart from "incitement to violence or hatred", such as "dissemination or distribution of materials", "condoning", "denying" or "trivialising crimes of genocide, crimes against humanity and war crimes", "instigating", "aiding" and even "abetting" in the commission of hate speech offences. Nevertheless, it does not broaden the scope of the target groups, as it focuses on racial and religious hatred.

Following the recommendations of the European Commission against Racism and Intolerance (ECRI), the Council of Europe defines hate speech as

> the advocacy, promotion or incitement, in any form, of the denigration, hatred or vilification of a person or group of persons, as well as any harassment, insult, negative stereotyping, stigmatisation or threat in respect of such a person or group of persons and the justification of all the primary types of expression, on the ground of race, colour, descent, national or ethnic origin, age, disability, language, religion or belief, sex, gender, gender identity, sexual orientation and other personal characteristics or status.[29]

the 607th Meeting of the Ministers' Deputies. https://rm.coe.int/CoERMPublicCommonSearchServices/DisplayDCTMContent?documentId=0900001680505d5b (accessed 27 May 2020).

28 Council Framework Decision 2008/913/JHA of 28 November 2008. Entry into force 6 December 2008. *Official Journal* OJ L 328 of 6 December 2008. This decision ensures that specific severe manifestations of racism and xenophobia are punishable by effective, proportionate and dissuasive criminal penalties throughout the European Union.

29 Council of Europe. No Hate Speech Movement. https://www.coe.int/en/web/venice/hate-speech-movement (accessed 27 May 2020).

In the definition above, one can find two core elements common to definitions of hate speech: public and wilful incitement to hatred directed against a legally-protected group of people. However, the definition is ambiguous for various reasons. Firstly, the use of the disjunctive conjunction "or" makes legal interpretation difficult because the options are presented as mutually exclusive when, in effect, they may not be exclusive. Secondly, the definition refers to a wide range of acts – e.g. "advocacy", "promotion", "incitement", justification", "harassment", "insult", "threat", and perlocutionary effects – e.g. "denigration", "vilification", "negative stereotyping" and "stigmatisation". Thirdly, the content points to a broad spectrum of legally-protected characteristics: "race", "colour", "descent", "national or ethnic origin", "age", "disability", "language", "religion", "belief", "sex", "gender", "gender identity" and "sexual orientation". Fourthly, the definition contains vague references, such as "other personal characteristics or status", that, in practice, would be difficult for legal practitioners to elucidate.

5.3.2 Member State law

This section illustrates different approaches to a technical legal definition of hate speech in Member State law. Due to spatial constraints, the analysis focuses on three EU Member States representing various legal cultures within European civil law: German civil law, French civil law and Spanish civil law.

a) Hate speech in German legislation

Although German legislation does not have a specific hate speech law, there are several penal, civil and administrative provisions against the most severe hate speech and hate crimes. The primary piece of criminal legislation prohibiting hate speech in Germany is the Criminal Code. This law provides a distinction between bias-motivated crimes (*Voruteilsdelikte*) and symbolic crimes (*Botschaftsverbrechen*). With respect to bias-motivated crimes, section 46, paragraph 2), of the Criminal Code states that

> racist, xenophobic and other inhumane or contemptuous motives are aggravating circumstances to be considered when establishing the grounds for sentencing for any crime under the Criminal Code.[30]

30 Criminal Code in the version published on 13 November 1998 (Federal Law Gazette I, p. 3322), as last amended by Article 2 of the Act of 19 June 2019 (Federal Law Gazette I, p. 844). Section 46. The German Federal Ministry of Justice provides a translation into English of the official German texts on their website. https://www.gesetze-im-internet.de/englisch_stgb/englisch_stgb.html#p1333 (accessed 29 May 2020).

In this respect, several provisions in the German Criminal Code are applicable in legal cases involving instances of hate speech, such as insult (under section 185), malicious gossip (under section 186), defamation (under section 187), malicious gossip and defamation concerning persons in political life (under section 188) and defiling the memory of the dead (under section 189). These provisions are applicable in both group defamation and defamation against a member of the group, in which there is bias motivation – that is, when the offences in question target a group or an individual that belongs to a group that can be identified by legally-protected characteristics.

On the other hand, symbolic crimes in Germany explicitly include “incitement to hatred”. Specifically, in section 130, Incitement of masses, of the Criminal Code, one can find two elements of hate speech: (1) speech that relates to incitement to hatred and (2) speech that is directed against people who belong to a legally-protected group. The German Criminal Code also describes the forms of expression of hate speech. These include: “incitement to hatred and violence”, “violation of the human dignity of others by disseminating, insulting, dehumanising and defaming”. In sum, German law seems to favour the right to dignity over the right to freedom of opinion and expression when the exercise of the latter is considered abusive.

b) Hate speech in French legislation

Hate speech in France is prohibited under both criminal law and civil law. The French Penal Code forbids any private defamation of a person or group of people for belonging to a group that can be identified by legally-protected characteristics (Article 132–176):

> The aggravating circumstances defined in the first paragraph are applicable when the offence is preceded, accompanied or followed by written or spoken words, images, objects or actions of whatever nature that damage the honour or the reputation of the victim, or a group of persons to which the victim belongs, on account of their actual or supposed membership or non-membership of a given ethnic group, nation, race or religion.[31]

31 Penal Code. English translation. With the participation of John Rason Spencer QC, Professor of Law, University of Cambridge, Fellow of Selwyn College. Article 132–76. Updated 12 October 2005, 49/132. file:///C:/Users/Victoria%20Guillen/Downloads/Code_34%20(2).pdf (accessed 29 May 2020).

Apart from the criminal provisions in the Penal Code, Article 23 of the Law on the Freedom of the Press of 29 July 1881 (as of 2014)[32] prohibits discrimination based on origin, membership, or non-membership of a racial or religious group. Moreover, Articles 32– 33 in the same law prohibit anyone from defaming or insulting a person or group identified by legally-protected characteristics. In addition, Article 34 prohibits defiling the memory of the dead. Therefore, in French legislation, hate speech is speech that defames, insults, or defames the memory of members of legally-protected groups of people based on ethnicity, nationality, race and religion. More recently, the categories of sex, sexual orientation and disability have been added to the already existing legally-protected characteristics.

c) Hate speech in Spanish legislation

As in the case of Germany and France, Spain has no specific hate speech law, nor does it have a specific legal definition of hate speech. However, there are penal, civil and administrative provisions that protect citizens from malicious acts of discrimination associated with hate speech. To begin with, at the state level, the Spanish Constitution[33] is based on principles of democratic values and non-discrimination on the basis of race, sex, religion, opinion or any other personal or social condition or circumstance. Specifically, Part I sets forth the individual's fundamental rights and duties, such as human dignity (under Article 10) and equality before the law (under Article 14). In addition, the rights to honour, personal and family privacy and to one's image are guaranteed (under Article 18).

The Spanish Penal Code[34] is considered the primary legislation that prohibits criminal offences associated with hate speech such as threats (under Article 169), defamation – calumny (under Articles 205 to 207) and injury (under Articles 208–210) – and attacks on honour and moral dignity (under Article 173). Furthermore, committing these criminal offences against legally-protected groups of people is considered an aggravating factor (under Article 170), which results in increased penalties for the offenders. In 2015, following the ECRI General Policy Recommendation No. 7 (revised) on national legislation to combat racism and

32 Loi du 29 Juillet 1881 sur la liberté de la presse. Version consolidée au 08 Juillet 2014. https://www.legifrance.gouv.fr/affichTexte.do;jsessionid=2D9E6AEE9BE04576DF46A63A4C088694.tpdjo05v_2?cidTexte=LEGITEXT000006070722&dateTexte=20140708 (accessed 29 May 2020).

33 Spanish Constitution 1978. https://www.boe.es/legislacion/documentos/ConstitucionINGLES.pdf Original title: Constitución Española 1978. BOE-A-1978.31229. https://www.boe.es/eli/es/c/1978/12/27/(1)/con (accessed 30 May 2020).

34 Organic Law No. 1/2015 of 30 March 2015, on Amendments to the Penal Code (Organic Law No. 10/1995 of 23 November 1995). http://melaproject.org/sites/default/files/2018-02/Spanish%20Criminal%20Code%20-%20Article%20510_0.pdf (accessed 30 May 2020).

racial discrimination,[35] the Spanish Penal Code was reformed. The most relevant improvement was that the essential elements of hate speech were included in the reformulation of Article 510, section 1, paragraph a) refers to the content, the purpose and the targets of hate speech:

> Those who, directly or indirectly, foster, promote or incite hatred, hostility, discrimination or violence against a group, or part thereof, or against a certain person for belonging to such a group, for reasons of racism, anti-Semitism or for other reasons related to ideology, religion or beliefs, family circumstances, the fact that the members belong to an ethnicity, race or nation, national origin, gender, sexual orientation or identity, or due to gender, illness or disability (Criminal Code 2016: 214).[36]

The reference made to hate speech in Article 510 is ambiguous and equivocal for various reasons. First, the speech acts constituting hate speech are presented as mutually exclusive options through the disjunctive conjunction "or" when in fact they may be concurrent (see Chapter 6 on complex speech acts) – e.g. "encourage", "promote" or "incite". Second, the various discourse strategies are presented as mutually exclusive when they may be not so – e.g. "directly" or "indirectly". Third, the effects on the targets are too broadly defined, as "discrimination" is also added to "hatred", "hostility" or "violence". Fourth, the legally-protected characteristics are wide-ranging – e.g. "or other grounds relating to ideology, religion or belief, family status, membership of an ethnic group, race or nation, national origin, gender, sexual orientation or identity, illness or disability". In paragraph b), Article 510 provides an exhaustive definition of dissemination. Nevertheless, the use of the disjunctive conjunction "or" hinders legal interpretation, as shown below:

> Those who produce, prepare, possess with the purpose of distributing, provide third parties access to, distribute, publish or sell documents or any other type of material or medium that, due to the content thereof, are liable to directly or indirectly foster, promote or incite hatred, hostility, discrimination or violence against a group, or part thereof, or against a certain person for belonging to such a group, for reasons of racism, anti-Semitism or for other reasons related to ideology, religion or beliefs, family circumstances, the fact that the

---

35 ECRI General Policy Recommendation No. 7 (revised) on National Legislation to Combat Racism and Rational Discrimination (adopted on 13 December 2002 and revised on 7 December 2017). https://www.coe.int/en/web/european-commission-against-racism-and-intolerance/recommendation-no.7 (accessed 31 May 2020).

36 Criminal Code. 2016. Ministry of Justice. Technical Secretariat (eds.). Colección Traducciones del Derecho Español. Translated by Clinter Traducciones e Interpretaciones, S.A. Madrid. https://www.mjusticia.gob.es/es/AreaTematica/DocumentacionPublicaciones/Documents/Criminal_Code_2016.pdf.

> members belong to an ethnicity, race or nation, national origin, gender, sexual orientation or identity, or due to gender, illness or disability [. . .] (Criminal Code 2016: 214).[37]

Moreover, in paragraph c), Article 510 refers to other malicious acts relating to hate speech, such as publicly denying, seriously trivialising or glorifying the crime of genocide; crimes against humanity or crimes against persons and property protected in the event of armed conflict, or the glorification of the perpetrators of such crimes. In section 2, paragraph a), Article 510 prohibits damage to human dignity by any means of public expression and dissemination. Paragraph b) penalises exaltation or justification of offences committed by another group against the target groups by any means of public expression or dissemination.

Another crucial legislative reference is Law 19/2007 against Violence, Racism, Xenophobia and Intolerance in Sport,[38] which is also ambiguous and difficult for legal practitioners to interpret because it is too broad in terms of (a) the speech acts – e.g. it prohibits conduct intended to "threaten", "intimidate", "insult", "humiliate" and "harass"; (b) the targets – e.g. individuals of vulnerable groups identified by legally-protected characteristics, such as "race", "ethnic", "geographical or social origin", "religion", "belief", "disability", or "sexual orientation"; and (c) the unlawful conduct -e.g. statements, gestures, singing, sounds, slogans and the display of banners, flags and symbols, amongst other instances.

6 Conclusions

In this chapter, Wittgenstein's concept of family resemblance (Wittgenstein 2009 [1953]) has helped us understand that hate speech does not have a single meaning but a family of meanings. Ordinary language analysis showed the family of meanings the term hate speech has in non-technical, everyday language. The three legal scholarly approaches to defining hate speech – content-based, intent-based and harms-based – ended up in a family of legal scholar definitions of hate speech, as did the approaches to a technical legal definition of hate speech in international law, common law and civil law – European Union law and Member State law.

---

37 Ibid, 214.

38 Law 18/2007, Against Violence, Racism, Xenophobia and Intolerance in Sport. https://www.global-regulation.com/translation/spain/1445414/law-19-2007%252c-of-july-11%252c-against-violence%252c-racism%252c-xenophobia-and-intolerance-in-sport.html Original title: Ley 19/2007, de 11 de julio, contra la Violencia, el Racismo, la Xenofobia y la Intolerancia en el Deporte. https://www.boe.es/buscar/pdf/2007/BOE-A-2007-13408-consolidado.pdf (accessed 30 May 2020).

Throughout this chapter, I have highlighted eight elements that can help define hate speech rigorously. I concur with Sellars (2016) in that these elements are: (1) the targeting of a group that is identifiable by legally-protected characteristics; (2) the content of the message only expresses hatred; (3) the speaker intends to harm or to encourage harmful activity; (4) the speech causes harm; (5) the speech is likely to incite wrong actions beyond the speech itself; (6) the speech is delivered in public; (7) the context makes violent response possible due to, for example, the power of the speaker, the receptiveness of the audience and the history of violence in the area where the speech is delivered; and (8) the speech has no other redeeming purpose.

I conclude that hate speech is an umbrella term that refers to negative forms of conduct intended to publicly incite hatred and violence against groups identifiable by legally-protected characteristics. Hate speech, a type of negative social behaviour, is likely to harm the dignity and equality of the target groups, keep them marginalised from mainstream society and ultimately destroy social cohesion and peace.

2 Hate Speech as a legal problem

1 Introduction

Hate speech as a legal problem is analysed in depth in the Report of the United Nations High Commissioner for Human Rights on the Expert Workshops on the Prohibition of Incitement to National, Racial or Religious Hatred, whose purpose is

> to provide a comprehensive assessment of the implementation of legislation, jurisprudence and policies regarding advocacy of national, racial or religious hatred that constitutes incitement to discrimination, hostility and violence at the national and regional levels while encouraging full respect for the fundamental right of freedom of opinion and expression as protected by international human rights law (Report of the United Nations 2013: 1).[39]

The Report of the United Nations points to some problematic aspects of hate speech legislation, such as, for instance, the existence of heterogeneous hate speech laws, ranging from excessively narrow to overly broad; the uneven and ad hoc applications of hate speech laws; the scarce jurisprudence on hate speech; the difficult task of adequately balancing the right to freedom of opinion and expression with the prohibition of incitement to hatred; the problems relating to curbing freedom of information and the use of the internet; and the difficulty of determining which speech acts can be defined as incitement to hatred and hence, prohibited by law, amongst others.

This chapter aims to elaborate on four significant issues that, in my view, are at the core of hate speech as a legal problem. These are: (a) the uneasy balance between freedom of expression and prohibition of incitement to hatred, (b) the lack of an agreed-upon technical legal definition, (c) the difficulty of determining which speech acts can be defined as incitement to hatred and (d) the legal challenges posed by online hate speech.

39 https://www.ohchr.org/Documents/Issues/Opinion/SeminarRabat/Rabat_draft_outcome.pdf (accessed 10 August 2020).

https://doi.org/10.1515/9783110672619-002

2 The uneasy balance between freedom of expression and the prohibition of incitement to hatred

According to Article 19 of the Universal Declaration of Human Rights:

> Everyone has the right to freedom of opinion and expression; this right includes the freedom to hold opinions without interference and seek, receive and impart information and ideas through any media and regardless of frontiers.[40]

According to Article 19, people have the right to think, speak freely and spread thoughts, ideas and opinions, whether good or bad, in writing or by any other means of expression. Therefore, legal decisions on the prohibition of hate speech need to be carefully balanced with the fundamental right to freedom of opinion and expression. Specifically, hate speech prohibition raises the question: Where should the legislator draw the line between free expression and hate speech? This question invites heated controversy. Some legislators are against regulating hate speech in any way because it restricts the fundamental right to freedom of opinion and expression, something which is considered a precondition for enjoying other human rights. Other legislators are in favour of regulating hate speech because it infringes the fundamental right to dignity (Article 1 of the Universal Declaration of Human rights) and implies an abusive exercise of freedom of expression (Article 19 of the Universal Declaration of Human Rights). Hate speech also arguably restricts the targets' right to freedom of opinion and expression. In the following sections, I will refer to the differing responses given in international law, United States constitutional law and European Union law concerning the balance between freedom of expression and hate speech prohibition.

2.1 International law

The United Nations High Commissioner for Human Rights argues in the Rabat Plan of Action (2013)[41] that when weighing freedom of expression against hate speech prohibition, states need to consider the following relevant issues:

---

40 The Universal Declaration of Human Rights (UDHR). The declaration was proclaimed by the United Nations General Assembly in Paris on 10 December 1948 (General Assembly resolution 217 A). https://www.ohchr.org/EN/UDHR/Documents/UDHR_Translations/eng.pdf (accessed 29 July 2020).

41 Report of the United Nations High Commissioner for Human Rights on the Expert Workshops on the Prohibition of Incitement to National, Racial or Religious Hatred. The Rabat Plan

1) Limitations of freedom of expression should protect a legitimate interest. In other words, limitations must be proportionate, so that the benefit to the protected interest outweighs the harm to freedom of expression.
2) Limitations of freedom of expression should protect the target groups from discrimination, hostility and violence rather than protect belief systems, religions or institutions from criticism.
3) Limitations of freedom of expression must be clearly and narrowly defined so that they do not restrict speech in a broad and untargeted way.
4) Only the most severe forms of offence should be criminalised. In this regard, it is essential to make a careful distinction between (a) forms of expression that should constitute a criminal offence, (b) forms of expression that are not criminally punishable but may justify a civil tort and (c) forms of expression that do not give rise to criminal or civil sanctions but should raise social concern in terms of tolerance and respect for the convictions of other people.

2.2 The United States constitutional law

The First Amendment to the Constitution of the United States protects the right to freedom of opinion and expression in the following terms:

> Congress shall make no law respecting an establishment of religion, or prohibiting the free exercise thereof; or abridging the freedom of expression, or of the press, or the right of the people peaceably to assemble and to petition the government for a redress of grievances.[42]

The First Amendment explains that freedom of expression is not absolute at all times and under all circumstances, as follows:

> There are certain well-defined and narrowly limited classes of speech, the prevention and punishment of which have never been thought to raise any constitutional problem. These include the lewd and obscene, the profane, the libellous, and the insulting or fighting words – those which by their very utterance inflict injury or incite an immediate breach of the peace.[43]

---

of Action on the Prohibition of Advocacy of National, Racial or Religious Hatred that Constitutes Incitement to Discrimination, Hostility or Violence. https://www.ohchr.org/EN/Issues/FreedomReligion/Pages/RabatPlanOfAction.aspx (accessed 10 August 2020).

42 The Constitution of the United States. First Amendment. https://constitution.congress.gov/constitution/amendment-1/ (accessed 27 May 2020).

43 Fighting Words, Hostile Audiences and True Threats: Overview. https://www.law.cornell.edu/constitution-conan/amendment-1/fighting-words-hostile-audiences-and-true-threats-overview (accessed 15 October 2022).

The Constitution of the United States requires a very high standard for restricting freedom of expression. Specifically, an individual's speech may be restricted under the United States constitutional law if it meets the imminence standard consisting of three criteria: (a) it is intended to produce lawless action, (b) it is likely to incite such action and (c) such action will likely incite an immediate breach of the peace. It is important to note that these criteria, especially (b) and (c), imply a very high standard for a court of justice to meet. In addition, these three criteria may be extremely problematic because they are bound to the external circumstances of each case. Instances of hate speech should always be analysed in their specific context of production, which may not always be easy for legal practitioners to access. According to the Constitution of the United States, speech restraints must remain an exception and be applicable only in severe cases involving imminent violence against the targets.

2.3 European Union law

Within the framework of European Union law, the ECRI General Policy Recommendation No. 15 on Combating Hate Speech (2015) states that

> Hate speech is based on the unjustified assumption that a person or a group of persons are superior to others; it incites acts of violence or discrimination, thus undermining respect for minority groups and damaging social cohesion. In this recommendation, ECRI calls for speedy reactions by public figures to hate speech; promotion of self-regulation of media; raising awareness of the dangerous consequences of hate speech; withdrawal of financial and other support from political parties that actively use hate speech; and criminalising its most extreme manifestations, while respecting freedom of expression. Anti-hate speech measures must be well-founded, proportionate, non-discriminatory, and not be misused to curb freedom of expression or assembly or suppress criticism of official policies, political opposition and religious beliefs.[44]

The ECRI General Policy Recommendation No. 15 discusses how far the right to freedom of opinion and expression can go in a democratic society (Article 1 of the Universal Declaration of Human Rights). The ECRI's recommendation also imposes a high standard for limiting freedom of speech, but not as high as the one set by the United States constitutional law. First, the law should only criminalise extreme manifestations of hate speech, those inciting violent acts against the target groups. Second, hate speech prohibition should not be misused by

44 European Commission against Racism and Intolerance (ECRI). General Policy Recommendation No. 15 on Combating Hate Speech. https://www.coe.int/en/web/european-commission-against-racism-and-intolerance/recommendation-no.15 (accessed 10 August 2020).

those in power to curb dissent and criticism. Third, freedom of expression should be extended to expressing ideas that contradict, attack, or upset the State, the democratic system, or a population group.

3 The lack of an agreed-upon technical legal definition of hate speech

As I pointed out in Chapter 1, at present, there is no agreed-upon technical legal definition of hate speech. Because of the lack of consensus about the legal meaning of hate speech, some people might argue that any attempt to regulate hate speech should be abandoned because regulating a legal concept for which there is no single definition is impossible. Hate speech is an umbrella term that encompasses a family of legal definitions drawn from several legal cultures. However, each definition differs from the next in perspective while sharing at least one quality with another definition in the family (cf. Wittgenstein's concept of family resemblance).

Apart from the problem of the lack of an internationally agreed-upon technical definition of hate speech, some jurisdictions have no technical legal definition of hate speech at all. In my view, this may have two significant implications. First, if hate speech is not cast in a technical legal definition, it may not be possible for the court of justice to recognise which forms of expression are likely to fall under the legally sanctionable category of hate speech. Second, certain forms of hate speech may pass unnoticed by the court of justice. As a result, victims of legally unrecognised hate speech could lack either protection of their fundamental rights to dignity and equality or lack parity of protection in a given jurisdiction compared to the victims of legally recognised hate speech in another jurisdiction. To illustrate the implications that the lack of an agreed-upon technical legal definition of hate speech may have, let us take the case of the United States, already discussed in Chapter 1. At the federal level, the Hate Crime Statistic Act defines a hate crime as one that:

> manifests evidence of prejudice based on race, gender and gender identity, religion, disability, sexual orientation, or ethnicity, including where appropriate the crimes of murder, nonnegligent manslaughter; forcible rape; aggravated assault, simple assault, intimidation; arson; and destruction, damage or vandalism of property.[45]

45 Hate Crime Statistics Act, As Amended, 28 USC. § 534 § [Sec. 1]. https://ucr.fbi.gov/hate-crime/2017/resource-pages/hate-crime-statistics-act.pdf (accessed 27 May 2020).

US law, then, prohibits hate crimes directed against groups or individuals that belong to groups identifiable by legally-protected characteristics. However, the definition does not refer to any crime that may fall under hate speech, incitement to hatred, group defamation, racist insults, racial harassment and threats. The lack of official recognition of what hate speech is makes the phenomenon invisible to the eyes of the law. Because of this, it may be the case that in US law, certain forms of hate speech may not be restricted when those same forms of hate speech are restricted in European Union law. Consequently, I argue that victims of hate speech may not have parity of protection in the United States compared with those in Europe. The expansion of the internet has foregrounded the protection imbalance between hate speech victims in the United States and European jurisdictions. As Tsesis argued:

> new regulatory challenges more frequently arise because of the global reach of hate propaganda transmitted from the US, where it is legal, and streamed into countries, like France, where such communications are criminal offences (Tsesis 2009: 497).

A technical legal definition needs to be clearly and narrowly defined and be sufficiently specific to enable lawyers to understand the meaning of the relevant legal concepts and ultimately assess potential liability in specific cases. As for existing hate speech prohibitions, these show varying degrees of specificity, use different terms to refer to hate speech offences, and some may be said to invade the terrain of freedom of expression. By way of illustration, in what follows, I analyse the definition of hate speech prohibition at different levels.

3.1 International law

At the international level, the prohibition of incitement to hatred was first established in Article 4, section a, of the International Convention on the Elimination of All Forms of Racial Discrimination (1965) in the following terms:

> Shall declare an offence punishable by law all dissemination of ideas based on racial superiority or hatred, incitement to racial discrimination, as well as all acts of violence or incitement to such acts against any race or group of persons of another colour or ethnic origin, and also the provision of any assistance to racist activities, including the financing thereof.[46]

---

46 https://www.ohchr.org/EN/ProfessionalInterest/Pages/CERD.aspx (accessed 7 August 2020).

The prohibition of incitement to hatred, then, is narrowly defined and is consistent with Article 19 of the Universal Declaration of Human Rights[47] because the restrictions imposed on freedom of expression are meant to protect individuals and communities against racial hatred. Specifically, limitations of speech refer to "ideas based on racial superiority or hatred", "incitement to racial discrimination", "acts of violence", or "incitement to such acts".

At the international level, another essential reference concerning the prohibition of incitement to hatred is Article 20, paragraph 2), of the International Covenant on Civil and Political Rights (1966): "Any advocacy of national, racial or religious hatred that constitutes incitement to discrimination, hostility or violence shall be prohibited by law".[48] This prohibition is clearly and narrowly defined, and is consistent with Article 19 of the Universal Declaration of Human Rights and Article 19 of the same law because it serves a legitimate interest: to protect the social rights of individuals and communities against national, racial or religious hatred. Interestingly, the prohibition is expressed through a restricted modifying adjective clause: "that constitutes incitement to discrimination, hostility or violence". This relative clause restricts the sentence meaning. In other words, the law prohibits "Any advocacy of national, racial or religious hatred" if it "constitutes incitement to discrimination, hostility or violence". The International Covenant on Civil and Political Rights (1966), unlike the International Convention on the Elimination of All Forms of Racial Discrimination (1965), also protects individuals and communities against national and religious hatred.

After analysing the two essential references in international law concerning the prohibition of incitement to hatred, I will review the definitions given in a short sample of common law and civil law jurisdictions to illustrate some significant discrepancies.

3.2 Common law

In Canadian law, section 319 of the Criminal Code[49] explicitly sets out for both "prohibition of incitement to hatred" and "wilful promotion of hatred" against any group identifiable by legally-protected characteristics in the following terms:

---

47 https://www.ohchr.org/EN/UDHR/Documents/UDHR_Translations/eng.pdf (accessed 7 August 2020).

48 https://www.ohchr.org/en/professionalinterest/pages/ccpr.aspx (accessed 7 August 2020).

49 Criminal Code, R.S.C. 1985, c. C-46, s. 319(1). https://laws-lois.justice.gc.ca/eng/acts/c-46/section-319.html (accessed 27 May 2020).

(1) Everyone who, by communicating statements in any public place, incites hatred against any identifiable group where such incitement is likely to lead to a breach of the peace is guilty of (a) an indictable offence and is liable to imprisonment not exceeding two years, or (b) an offence punishable on summary conviction.
(2) Everyone who, by communicating statements other than in private conversation, wilfully promotes hatred against any identifiable group is guilty of (a) an indictable offence and is liable to imprisonment for a term not exceeding two years; or (b) an offence punishable on summary conviction.

The Canadian Criminal Code provides a clear and narrow definition that is consistent with Article 19 of the Universal Declaration of Human Rights (1948) and Article 19 of the International Covenant on Civil and Political Rights (1966) because it only restricts speech that constitutes incitement to hatred "where such incitement is likely to lead to a breach of the peace". Nevertheless, one might argue that the definition of the Canadian Criminal Code is too broad about the way perpetrators can "incite" or "wilfully promote hatred", as it merely states "by communicating statements". The definition also makes a vague reference to the locus – "in a public place" – and is overly general about the legally-protected groups in question – "any identifiable group".

In UK law, racial hatred is defined in section 17 of Part III Racial Hatred of the Public Order Act 1986 in the following terms: "racial hatred means hatred against a group of persons defined by reference to colour, race, nationality (including citizenship) or ethnic or national origin."[50] Under this definition, only incitement to racial hatred is considered an offence. Over the years, other groups identifiable by legally-protected characteristics have joined the Public Order Act 1986, specifically religious hatred joined in 2006 and hatred on sexual orientation grounds joined in 2008. Section 18 of Part III Racial Hatred of the Public Order Act (1986) sets out the acts intended or likely to stir up racial hatred:

Use of words or behaviour or display of written material

(1) A person who uses threatening, abusive or insulting words or behaviour or displays any written material which is threatening, abusive or insulting is guilty of an offence if -
    (a) he intends thereby to stir up racial hatred, or
    (b) having regard to all the circumstances, racial hatred is likely to be stirred up thereby.

50 The Public Order Act 1986. https://www.legislation.gov.uk/ukpga/1986/64 (accessed 24 July 2020).

Section 18 of Part III Racial Hatred of the Public Order Act (1986) requires a high standard because the law can only restrict speech if the court of justice finds that the speaker intends to stir up racial hatred.

In Australian legislation, section 77 of the Racial Vilification Act prohibits “racial animosity” or “racial harassment” in the following terms:

> Any person who engages in any conduct, otherwise than in private, by which the person intends to create, promote or increase animosity towards, or harassment of, a racial group or a person as a member of a racial group is guilty of a crime and is liable to imprisonment for 14 years.[51]

The Australian definition is a clear example of the heterogeneity of hate speech laws I noted above. To begin with, it does not include any of the key terms contained in Article 20, paragraph 2), of the International Covenant on Civil and Political Rights (1966) already noted, which is considered an international standard with the purpose of guiding domestic legal frameworks on incitement to hatred. The key terms the definition should but does not include are “incitement to hatred”, “hostility” and “violence”. Secondly, it uses different terminology from the other definitions I have looked at – e.g. “promote or increase animosity or harassment”. Thirdly, the definition is overly broad about the type of conduct that is likely to promote or increase animosity or harassment against the target group – e.g. “any conduct, otherwise than in private, by which the person intends to create, promote or increase animosity towards, or harassment of”. Lastly, the definition only protects the targets “from racial animosity or harassment”, setting aside other widely recognised characteristics such as “religion”, “sex”, “sexual orientation” and “disability”.

3.3 Civil law

The first example comes from European Union law. Following the recommendations of the European Commission against Racism and Intolerance (ECRI), the Council of Europe prohibits

> The advocacy, promotion or incitement, in any form, of the denigration, hatred or vilification of a person or group of persons, as well as any harassment, insult, negative stereotyping, stigmatisation or threat in respect of such a person or group of persons and the justification of all the primary types of expression, on the ground of race, colour, descent,

51 The Racial Vilification Act 2004. Section 77. https://www.legislation.wa.gov.au/legislation/statutes.nsf/law_a9290_currencies.html (accessed 28 May 2020).

> national or ethnic origin, age, disability, language, religion or belief, sex, gender, gender identity, sexual orientation and other personal characteristics or status.[52]

The Council of Europe's definition is an excellent example of an overly broad definition of hate speech. It is especially vague on the types of conduct that are to be prohibited – e.g. "The advocacy, promotion or incitement, in any form", and presents new legally-protected categories: "descent", "disability", "language", "belief", "sex", "gender" and "gender identity".

Our last example comes from Spanish civil law. Article 510 of the Criminal Code (2015) prohibits incitement to hatred in the following terms:

> Whoever publicly encourages, promotes or incites, directly or indirectly, hatred, hostility, discrimination or violence against a person as a part of a group or group on the grounds of racism, anti-Semitism or other grounds relating to ideology, religion or belief, family status, membership of an ethnic group, race or nation, national origin, gender, sexual orientation or identity, illness or disability (section 1, paragraph a).[53]

The influence of European Union law on the legal prohibition of incitement to hatred in Spanish law is evident. Article 510 even broadens the list of protected categories that the Council of Europe provides in its definition: "racism, anti-Semitism or for other reasons related to ideology, religion or beliefs, family circumstances, the fact that the members belong to an ethnicity, race or nation, national origin, gender, sexual orientation or identity, or due to gender, illness or disability" (Criminal Code 2016: 214). The broadness of Article 510 also affects the category or element of "dissemination", to the extent that it incorporates more limitations to freedom of opinion and expression:

> Those who produce, prepare, possess with the purpose of distributing, provide third parties access to, distribute, publish or sell documents or any other type of material or medium that, due to the content thereof, are liable to directly or indirectly foster, promote or incite hatred, hostility, discrimination or violence against a group [. . .] (section 1, paragraph b)) (Criminal Code 2016: 214).

Article 510, in paragraph c), also refers to other malicious acts that the legal concept of hate speech may encompass:

---

52 Council of Europe. No Hate Speech Movement. 2015. https://www.coe.int/en/web/venice/hate-speech-movement (accessed 27 May 2020).

53 All quotes from Article 510 have been borrowed from the official English translation, Criminal Code 2016, edited by the Ministry of Justice and Technical Secretariat and translated by Clinter Traducciones e Interpretaciones. https://www.mjusticia.gob.es/es/AreaTematica/DocumentacionPublicaciones/Documents/Criminal_Code_2016.pdf.

> Those who publicly deny, seriously trivialise or extol the crimes of genocide, crimes against humanity or against persons and property protected in the event of armed conflict, or who extol the perpetrators thereof, if committed against a group or part thereof, or against a certain person for belonging to such a group, for reasons of racism, anti-Semitism or for other reasons related to ideology, religion or beliefs [. . .] if such conduct promotes or encourages a climate of violence, hostility, hatred or discrimination against such individuals (section 1, paragraph c)) (Criminal Code 2016: 214).

In section 2, paragraph a), Article 510 prohibits damage to human dignity by any means of public expression and dissemination. Paragraph b) of the same section penalises exaltation or justification of offences committed by another group against the target groups by any means of public expression or dissemination. For many jurists in Spain (cf. Landa-Gorostiza 2018), the broadness of Article 510 may have significant implications. First, it may be inconsistent with Article 19 of the Universal Declaration of Human Rights (1948), Article 19 of the International Covenant on Civil and Political Rights (1966) and the ECRI General Policy recommendation No. 15 (2015). Second, it may give rise to controversial interpretations of the law. Third, it may invite differing ruling decisions in the same case or similar cases of hate speech.

4 The task of determining which speech acts can be defined as incitement to hatred

The Rabat Plan of Action proposes a threshold test to assess the forms of expression and speech acts that are likely to fall into the category of incitement to hatred. The threshold test consists of six criteria. In the following, I will refer to each of the mentioned criteria in further detail.

1) Context. The socio-political context in which hate speech was produced.
2) Speaker. The speaker's power to incite other people to commit violent acts against the target group.
3) Intent. Guilty mind (*mens rea*) – that is, whether the speaker wilfully intends to incite hatred against the target group.
4) Content and form – e.g. style, argumentation and directness.
5) The extent of the speech act. The magnitude of the speech act – e.g. whether the hateful messages are circulating in a restricted environment or are widely accessible to the general public, the means of dissemination used by the speaker, the quantity and frequency of the communications and the capacity of violent response of the audience.
6) Likelihood. The probability that the speech act would imminently succeed in inciting hatred, hostility or violence against the target group.

The pragmatic turn in law (Giltrow & Stein 2017) is visible in the above six criteria that legal practitioners bear in mind to decide which speech acts may fall into the category of hate speech.

5 The legal challenges raised by online hate speech

Since the internet came into wide use in the 1990s, it has become, according to Delgado and Stefancic, "the site of some of the worst forms of racial and sexual vituperation" (Delgado & Stefancic 2018: 43). The internet, in effect, disseminates and propagates hateful statements of all kinds, such as racist, anti-Muslim, antisemitic, misogynist and homophobic messages. The dissemination of hateful remarks and messages indeed existed before the internet, but the new medium has made such dissemination easier and less costly for hate-advocating speakers. The proliferation of online hate speech (Winter 2019; Lumsden & Harmer 2019; Udupa, Gagliardone & Hervik 2021; Banaji & Bhat 2022) has raised regulatory difficulties, such as the ones I refer to in the following:

a) Multiple authorship. More often than not, the production and dissemination of online hate speech involve multiple actors. When hate-advocating speakers use an online platform to disseminate their hateful messages, they hurt the victims and may violate terms of service on that platform and, at times, even State law, depending on their location. The victims, for their part, may feel helpless in the face of online hate speech because authorship is mostly unknown; hence they do not know against whom they could file a complaint.
b) Anonymity. The internet provides hate-advocating speakers with anonymity; hence, perpetrators have the chance to avoid responsibility for what they say. Besides, speakers feel freer to speak hate in anonymous or pseudo-anonymous speech because they know they cannot be charged with a criminal offence.
c) Audience. After hate-advocating speakers have delivered hateful statements over the internet, it is difficult to say how many readers will read the messages, when they will read them, and what effects the messages will have on the audience. Unlike hate words conveyed in a protest speech at a rally or printed on a pamphlet, online hate speech can be read worldwide, reread, printed, or pass unnoticed. The audience's reactions to online hate speech are highly unpredictable.
d) Permanence. Hate speech can be online for a long period, in different formats and across multiple platforms, linked repeatedly. The longer the hateful

messages stay available, the more damage they can inflict on the victims and hence, empower the hate-advocating speaker.

e) Itinerancy. Even if a service provider discovers misuse and cancels a user's account – which is, in practice, a difficult task considering the vast contents of the internet – hate speech may find expression elsewhere, possibly by the same speaker on the same platform under a different name, or on different online spaces.
f) Deception. The internet provides an inexpensive way for hate-advocating groups to spread their messages, using various methods of deception to lure internet users to their websites – e.g. email, instant messaging and chat rooms. Besides, hate speech may be hidden beneath layers of patriotic speech, cartoons and advertisements.
g) The transnational reach of the internet. Because of the cross-jurisdictional discrepancies in the prohibition of incitement to hatred, there is an urgent need for international cooperation concerning legal mechanisms for combating online hate speech.

Given the unique nature of online hate speech, it seems reasonable to think that alternative tests are needed to balance the competing interests of the right to freedom of opinion and expression with those of the right to dignity of the target groups in cyberspace (Tsesis 2009). For example, in the context of US law, traditional legal doctrines have proved to be inadequate to evaluate online hate speech. As Komasara argued, "the fighting words doctrine lacks the public, face-to-face element needed to increase the level of inciting an immediate breach of the peace" (Komarasa 2002: 844). Brenner drew attention to the fact that although the court's current standard under Brandenburg – imminence standard – may be suitable for traditional broadcast or print media, online hate speech has made such a standard outdated. For Brenner, the application of the Brandenburg test "makes it more difficult for a jury to convict the sender of an online hate message or remark because the connection between words broadcast over the internet and the reader's reaction is difficult to gauge" (Brenner 2002: 781).

In the same line of thought, Cronan claimed that the imminence standard does not work with online hate speech because "words in cyberspace are usually heard well after they are spoken. As a result, almost no internet communication, regardless of the likelihood and seriousness of incitement, can be condemned under Brandenburg" (Cronan 2002: 428). Therefore, the Brandenburg test seems to create an impossible hurdle to imposing liability on internet communications. The question is whether the imminence should be calculated from the time the message is posted (the speaker's perspective) or from the time the message is read (the recipient's perspective). In addition, the message may not always contain an explicit incitement to hatred but an implicit one[54] (see Chapters 6 and 8 on implicitness and indirectness).

At present, there is an ongoing legal debate about alternative tests to better balance free speech concerns with protecting potential victims against speech that "plants the seeds of hatred by combining information and incitement that ultimately enables others to commit violence" (Komasara 2002: 837). Several courts of appeal in the US have turned to the concept of *true threats*, an unprotected speech act by the First Amendment, into a basis for the hate-advocating speaker's civil liability or criminal prosecution. The task of determining what type of communication constitutes a true threat was addressed in the case United States v. Alkhabaz (1997):

> Accordingly, to achieve the intent of Congress, we hold that, to constitute "a communication containing a threat" under section 875(c), a communication must be such that a reasonable person (a) would take the statement as a serious expression of an intention to inflict bodily harm (the mens rea), and (b) would perceive such expression as being communicated to effect some change or achieve some goal through intimidation (the actus reus).[55]

To avoid hate-advocating speakers hiding under the First Amendment's protection mantel, legal scholars have proposed various approaches. Cronan suggested a formulation for an internet incitement standard that meets the specific demands of cyberspace but remains faithful to the imminence standard. In his view, the internet incitement standard should consider four essential factors: (a) imminence from the perspective of the recipient, (b) message content, (c) likely audience and (d) nature of the issue involved (Cronan 2002: 455). In practice, factors (a) and (b) may be difficult for the courts to elucidate.

Delgado (1982) suggested a new no-imminence standard according to which hate-advocating internet users should bear some civil liability if a causal link can be found between their hate advocacy and the harm inflicted on the target groups. Specifically, Delgado proposed regulating online hate speech through the torts of intentional infliction of emotional distress or group defamation. In his view, tort law can supply models for harms-based codes compatible with existing restrictions on the right to freedom of opinion and expression under the First Amendment to the Constitution of the United States.

---

54 Significantly, at the heart of the so-called January 6 hearings, whose main purpose is to publicly present evidence of President Donald Trump's culpability for the January 6 Capitol assault, is whether or not Trump's statements contained implicit incitement to violence.

55 US Court of Appeals. United States v. Alkhabaz. 1997. No. 104 F.3d 1492. https://law.justia.com/cases/federal/appellate-courts/F3/104/1492/549096/ (accessed 6 May 2021).

6 Conclusions

This chapter analysed four essential issues concerning hate speech as a legal problem: (a) the uneasy balance between the right to freedom of opinion and expression and the prohibition of incitement to hatred, (b) the lack of an agreed-upon technical legal definition, (c) the difficulty of determining which speech acts can fall into the category of incitement to hatred and (d) the legal challenges raised by online hate speech.

One can learn from this chapter that weighing the right to freedom of opinion and expression against the prohibition of incitement to hatred is no simple task. International human rights standards, whose aim is to guide legislation at the domestic level, set a high standard to limit freedom of opinion and expression. For this reason, the prohibition of incitement to hatred should be clearly and narrowly defined. The chapter also drew attention to the fact that not all domestic legislations follow the international law recommendations accurately. There are, in effect, troublesome discrepancies between anti-incitement laws in countries worldwide. The definitions these laws provide concerning the prohibition of incitement to hatred range from too narrow to overly broad.

Another significant aspect of hate speech as a legal problem is deciding which speech acts can fall into the category of incitement to hatred. Although it would be useful for legal practitioners to have a closed catalogue or inventory of hate speech acts and forms of expression, the Rabat Plan of Action (2015) made it clear that this is an impossible task, given that the analysis of hate speech requires the analysis of the context in which hate speech was produced. It is noteworthy that the approach recommended by the Rabat Plan of Action is pragmatics-based. In European Union hate speech law, it is the social context together with the speaker's communicative intent (illocutionary act), rather than merely the words the speaker says (locutionary act), and the potentially harmful effects the message may have upon the target groups (perlocutionary act), that are essential elements for the court of justice to make a legal decision (see Chapter 6 on Speech act theory).

Furthermore, online hate speech has raised new challenges for legal practitioners. Because of the internet's global reach, online hate speech may cross transnational jurisdictions, jurisdictions which may not always share similar views on the limits of freedom of opinion and expression. New historical conditions and constantly evolving means of communication call for new approaches to balancing the right to freedom of opinion and expression against the right to dignity. In the case of US law, the Supreme Court's current imminence standard has proved to be an inadequate legal tool for curbing online hate speech.

3 The legal reasoning in hate speech court proceedings

1 Introduction

As a consequence of the Holocaust against Jews (1941–1945), hate speech prohibition was placed at the heart of modern international human rights. A look at the Universal Declaration of Human Rights (1948), the International Convention on the Elimination of All Forms of Racial Discrimination (1965) and the International Covenant on Civil and Political Rights (1966), to name but a few key universal standards in the field of international law, shows a heated debate on the extent to which the right to freedom of opinion and expression should provide its protective mantle to speech that is used to destroy the rights to dignity and equality of other people (cf. Delgado & Stefancic 2018; Tsesis 2009). Although international human rights standards drive democratic societies worldwide, hate speech prohibition has developed differently on both sides of the Atlantic, as discussed in Chapter 2.

This chapter analyses the key foci of hate speech legal reasoning across different jurisdictions. Due to spatial constraints, the scope of the analysis is reduced to the United States and the European Union, as they provide representative examples of different approaches to hate speech in common law and civil law jurisdictions. The selection of landmark cases from different decades is intended to indicate some permanent and long-lasting issues, such as the balance the courts of justice on both sides of the Atlantic must find between the competing interests of the right to freedom of opinion and expression and the right to dignity, and more particularly the difficulties that face legal practitioners in applying the imminence standard in US law. Besides, since the legal cases under discussion all reached the higher courts of justice, they also illustrate the lack of consistency in hate speech legal reasoning and decisions across jurisdictions. The chapter begins by analysing several landmark decisions of the United States Supreme Court and the United States Court of Appeals for the Armed Forces. Then, some significant legal decisions of the European Court of Human Rights (ECHR) are discussed. Finally, the chapter contrasts the key foci of legal reasoning in hate speech cases in the United States with those in the European Union.

https://doi.org/10.1515/9783110672619-003

2 The United States Supreme Court

The United States Supreme Court is considered the gatekeeper of the Federal Constitution. It is the highest court in the land and, for this reason, the court of last resort for those seeking justice. The Supreme Court plays an essential role in protecting civil rights and liberties by striking down State laws that violate the Constitution. In addition, it ensures that the changing views of a majority do not undermine the fundamental values common to all Americans – that is, freedom of speech, freedom of religion and due process of law.

In the following sections, I analyse some landmark decisions[56] of the United States Supreme Court and the United States Court of Appeals for the Armed Forces, to learn how the courts balance the right to freedom of expression under the First Amendment and other fundamental rights when deliberating on hate speech cases.

2.1 The case of Terminiello v. Chicago (1949)

Terminiello v. Chicago (1949)[57] is a landmark court case[58] in US jurisprudence in which the United States Supreme Court held that a breach of the peace ordinance of the city of Chicago was unconstitutional under the First Amendment. The case grew out of a speech by Arthur Terminiello, a Catholic priest under suspension and an anti-Semite, who railed against Jews and communists in an auditorium in Chicago under the auspices of the Christian Veterans of America. By way of illustration, I quote two excerpts from his inflammatory speech, excerpts which are included in the Supreme Court's judgment:

> It is the same kind of tolerance if we said there was a bedbug in bed, "We don't care for you" or if we looked under the bed and found a snake and said, "I am going to be tolerant and leave the snake there." We will not be tolerant of that mob out there. We are not

---

56 A landmark decision is a decision that is often cited because it changes, consolidates, updates, or summarises the law on a particular topic.

57 US Supreme Court. Terminiello v. Chicago, 337 US 1 (1949). Terminiello v. Chicago. No. 272. Argued 1 February 1949. Decided 16 May 1949. https://supreme.justia.com/cases/federal/us/337/1/ (accessed 18 August 2020).

58 A landmark case is a court case that has historical and legal significance. The study of landmark cases improves understanding of how past judicial decisions have affected the law and how they will be applied to current cases. US law is a common law system in which judges base their legal decisions on previous court rulings in similar cases. Consequently, previous legal decisions by a higher court are legally-binding and become part of case law.

> going to be tolerant any longer. So, my friends, since we spent much time tonight trying to quiet the howling mob, I am going to bring my thoughts to a conclusion, and the conclusion is this. We must all be like the Apostles before the coming of the Holy Ghost. We must not lock ourselves in an upper room for fear of the Jews. I speak of the Communistic Zionistic Jew, and those are not American Jews. We don't want them here; we want them to go back where they came from (Terminiello v. Chicago, 337 US 1 (1949), p. 337 U.S. 21).

While the auditorium was full, a furious crowd of about one thousand people had gathered outside in order to picket and protest against the meeting. Although a cordon of police officers tried hard to maintain order, they could not prevent a riot. Terminiello's speech inflamed the crowd, and fights broke between the audience members and the protesters outside. As a result, the police arrested Terminiello for riotous speech. He was judged and found guilty of violating Chicago's breach of the peace ordinance. This ordinance banned speech which stirs the public to anger, invites a dispute, or creates a disturbance. Terminiello appealed the local court's decision, claiming that the ordinance violated his right to freedom of expression under the First Amendment. The Illinois Appellate Court and the Illinois Supreme Court upheld the guilty verdict. Nevertheless, the United States Supreme Court overturned the conviction on the grounds of the First Amendment.

At the heart of the legal discussion were two critical issues for the Supreme Court to deliberate. The first issue was whether Chicago's breach of the peace ordinance was unconstitutional. The second issue was whether Terminiello's right to freedom of expression had been violated in the case or, on the contrary, he had abused this right by uttering fighting words – that is to say, words that, by their very utterance, inflict injury or tend to incite violence and hence are not protected by the First Amendment.

The majority opinion reversed Terminiello's conviction, holding that the First Amendment protected his speech and that the ordinance, as construed by the Illinois courts, was unconstitutional because it invaded the province of freedom of speech. The majority opinion explained that the function of free speech is to invite dispute, even when it induces a condition of unrest or even stirs people to anger. The majority opinion further argued that the right to speak freely and to promote diversity of ideas is, therefore, one of the chief distinctions that differentiate democratic societies from totalitarian regimes.

By contrast, the dissenting opinion drew attention to the fact that Terminiello had abused his right to freedom of expression by uttering fighting words. For instance, Terminiello vilified Jews by calling them "slimy scum", "dirty kikes", "bedbugs" and "snakes", while he praised fascists leaders. For the dissenting opinion, Terminiello's speech comprised utterances that are not an essential part of any exposition of ideas. The speech contained anti-Jewish stories expressed through inflammatory language that incited hatred and violence. For the dissenting opinion, reversing Terminiello's conviction implied a dogma of absolute freedom for irresponsible utterances so provocative that they even challenged the power of local authorities to keep social order and peace. Justice Jackson's dissent, in this case, is most famous for its final paragraph, which summarises the controversial debate around the need to set limits on the right to freedom of expression if individuals abuse it:

> This Court has gone far towards accepting the doctrine that civil liberty means the removal of all restraints from these crowds and that all local attempts to maintain order are impairments of the citizen's liberty. The choice is not between order and liberty. It is between liberty with order and anarchy without either. There is a danger that if the Court does not temper its doctrinaire logic with a little practical wisdom, it will convert the constitutional Bill of Rights into a suicide pact (Terminiello v. Chicago, 337 US 1 (1949), p. 337 U.S. 37).

2.2 The case of Brandenburg v. Ohio (1969)

Brandenburg v. Ohio (1969)[59] is another landmark case in US jurisprudence in which the Supreme Court established the imminence standard, also known as the Brandenburg test, for speech that advocates violence (see Chapter 2). The case is considered most significant because of its ongoing effect on the application of the law in cases associated with hate speech. In Brandenburg v. Ohio (1969), the imminence standard superseded the clear and present danger standard (Schenck v. the United States 1919),[60] which until then had been the doctrine used by the Supreme Court to determine under what circumstances limits can be placed on First Amendment freedoms of speech, press, or assembly. The distinct elements of the Brandenburg test are (1) intent to speak, (2) imminence of lawlessness and (3) likelihood of lawlessness. The imminence standard involves more protection to freedom of expression than the clear and present danger standard, because of the temporal restriction it enacts – that is, advocacy of violence can only be banned if it is intended to incite the commission of imminent lawless action, and it is likely to result in such lawless action.

59 US Supreme Court. Brandenburg v. Ohio, 395 US 4444 (1969). Brandenburg v. Ohio. No. 492. Argued 27 February 1969. Decided 9 June 1969. https://supreme.justia.com/cases/federal/us/395/444/ (accessed 18 August 2020).

60 US Supreme Court. Schenck v. the United States, 249 US, 47 (2019). Schenck v. the United States. No. 437, 438. Decided 3 March 1919. https://supreme.justia.com/cases/federal/us/249/47/ (accessed 20 August 2020).

The case record showed that Brandenburg, a Ku Klux Klan leader in rural Ohio, contacted a reporter at a Cincinnati television station and invited him to cover a Ku Klux Klan rally that would take place on a farm in Hamilton County. As a result, portions of the rally were broadcast on the local station and a national network. The prosecution's case rested on the footage and testimony identifying Brandenburg as the person who spoke with the reporter and at the rally. The footage showed objects such as firearms, ammunition and a Bible. Specifically, one scene of the first film showed twelve hooded figures, some of whom carried firearms. They were gathered around a large burning cross. Although most of the speech uttered during this scene was unintelligible, some statements could be distinctively understood. The speech contained derogatory terms directed at African Americans and the Jewish people:

> How far is the nigger going to – yeah.
> This is what we are going to do to the niggers.
> A dirty nigger.
> Send the Jews back to Israel.
> Let's give them back to the dark garden.
> Save America.
> Let's go back to constitutional betterment.
> Bury the niggers.
> We intend to do our part.
> Give us our state rights.
> Freedom for the whites.
> Nigger will have to fight for every inch he gets from now on (Brandenburg v. Ohio, 395 US 4444 (1969), p. 395 U.S. 449).

According to the case record, another scene in the first film showed Brandenburg, whose red hood identified him as a leader within the Ku Klux Klan, giving a speech. I here quote the speech excerpt examined by the Supreme Court:

> This is an organisers' meeting. We have had quite a few members here today which are – we have hundreds, hundreds of members throughout the State of Ohio. I can quote from a newspaper clipping from the Columbus, Ohio, Dispatch, five weeks ago Sunday morning. The Klan has more members in the State of Ohio than does any other organisation. We're not a revengent [sic] organisation, but if our President, our Congress, our Supreme Court, continues to suppress the white, Caucasian race, it's possible that there might have to be some revengeance taken (Brandenburg v. Ohio, 395 US 4444 (1969), p. U.S. 446).

The second film showed six hooded figures one of whom was Brandenburg repeating a speech very similar to the one in the first film, except for one significant difference. Whereas the possibility of "revengeance" was omitted, one sentence was added: "Personally, I believe the nigger should be returned to Africa, the Jew returned to Israel" (Brandenburg v. Ohio, 395 US 4444 (1969), p. US 447).

The speech in the first film explicitly referred to the possibility of "revengeance", an obsolete form of the word "vengeance", against African Americans, the Jewish people and those who supported them. In the context in which the speech was uttered, the sentence: "if our President, our Congress, our Supreme Court, continues to suppress the white, Caucasian race, it's possible that there might have to be some revengeance taken", could be interpreted as a threat because it states an intention to inflict injury on African Americans and the Jewish people as a response to the civil rights they have been given by US authorities. Brandenburg was charged under Ohio's Criminal Syndicalism Statute, banning the advocacy of violence or crime as a means of political reform, for his participation in the rally and for his inflammatory speech. On appeal, the Ohio First District Court of the United States of Appeal affirmed Brandenburg's conviction, rejecting his claim that the statute violated his right to freedom of expression. The Supreme Court of Ohio dismissed his appeal without opinion. However, the Supreme Court of the United States overturned Brandenburg's conviction. This controversial legal decision held that the local government could not constitutionally punish abstract advocacy of force or law violation unless directed to incite or produce imminent lawless action, and such advocacy is likely to produce or incite such action.

2.3 The case of National Socialist Party v. Skokie (1977)

The case of National Socialist Party v. Skokie (1977)[61] contains a landmark decision of the US Supreme Court related to freedom of opinion and expression and freedom of assembly. According to the case record, a leader of the National Socialist Party of America (NSPA) – a Neo-Nazi group – repeatedly held demonstrations in Marquette Park where the party had its headquarters. The Chicago authorities sought to block his plan to hold yet another of these white power demonstrations by requiring the NSPA to post a public safety insurance bond and by banning political demonstrations in Marquette Park. The NSPA leader then filed a claim against the city of Chicago for violating his rights under the First Amendment. In addition, the NSPA leader sent out letters to the park districts of the North Shore suburbs of Chicago requesting permits for the NSPA to hold a white power demonstration. Some suburbs decided to ignore the letters, while the village of Skokie, which was the home to many Holocaust survivors, decided to respond. The village of Skokie successfully filed for an injunction against the NSPA and passed several ordinances to prevent any future requests like the NSPA's. Whereas one of the

61 US Supreme Court. National Socialist Party v. Skokie (1977). No. 76–1786. Decided: 14 June 1977. https://caselaw.findlaw.com/us-supreme-court/432/43.html (accessed 18 August 2020).

ordinances stated that people could not wear Nazi uniforms or display swastikas during demonstrations, the two other ordinances banned distributing hate speech material. The ordinances ensured that demonstrations could be held, provided that the organisers paid for a substantial public safety insurance bond. In effect, the ordinances rendered it impossible for the NSPA to hold the demonstration.

The NSPA leader used both the injunction and the ordinances as an opportunity to claim infringement upon his First Amendment rights and subsequently demanded to protest in Skokie for the NSPA's right to freedom of expression. On behalf of the NSPA, the American Civil Liberties Union (ACLU) challenged the injunction on the grounds of infringement upon the First Amendment rights of the marchers. The ACLU appealed on behalf of the NSPA, but both the Illinois Appellate Court and the Illinois Supreme Court refused to stay the injunction.

The ACLU then appealed the refusal to the United States Supreme Court, which *granted certiorari.*[62] Skokie's attorneys argued that for the Holocaust survivors, seeing the display of swastikas was like receiving a physical attack. The Supreme Court rejected the argument, ruling that the display of swastikas is a symbolic form of free speech entitled to First Amendment protections and determined that displaying swastikas did not constitute an imminent threat. The Supreme Court's ruling allowed the NSPA to march. Furthermore, as a result of litigation in the federal courts (Collin v. Smith 1978),[63] the village's ordinance was declared unconstitutional for infringement upon First Amendment rights.

2.4 The case of Virginia v. Black (2003)

The case of Virginia v. Black (2003)[64] arises from the criminal conviction of three Ku Klux Klan members in two separate cases for violation of a statute of Virginia State outlawing the burning of crosses, either on the private property of another or in public places, with the intent to intimidate or place other

62 In the US Supreme Court, if four Justices agree to review the case, then the Court will hear the case. This act is referred to as *granting certiorari*. The legal term *certiorari* comes from Latin and means to be more fully informed. The term *certiorari* is also used in the legal expression *writ of certiorari*, which orders a lower court to deliver its record in a case so that the higher court may review it.

63 Collin v. Smith, 578 F.2d 1197 (7th Cir. 1978). Decided 22 May 1978. https://law.justia.com/cases/federal/appellate-courts/F2/578/1197/448646/ (accessed 18 August 2020).

64 US Supreme Court. Virginia v. Black, 538 US 343 (2003). Certiorari to the Supreme Court No. 01–1107. Decided 7 April 2003. https://supreme.justia.com/cases/federal/us/538/343/ (accessed 24 August 2020).

people in fear of bodily harm. In the first case, the subject attempted to light a cross on the property of a Black person. In the second case, the applicant held a Ku Klux Klan rally on private property. During the rally, derogatory terms were used against African Americans and a cross burned. Upon their convictions, each subject appealed to the Supreme Court of Virginia, contending that the statute was facially unconstitutional because it viewed the physical act of burning a cross as sufficient evidence of intimidation (*prima facie* evidence).[65] The Supreme Court of Virginia declared the statute unconstitutional because its *prima facie* evidence provision posed the risk of prosecuting the legitimate exercise of symbolic expressions and petitioned the decision to the United States Supreme Court. At the heart of the legal deliberation was the question of whether or not Virginia's statute violated the right to freedom of expression, because of its *prima facie* evidence provision. In the following, I summarise the most relevant aspects of the Supreme Court's legal reasoning.

On the one hand, the Supreme Court's majority opinion argued that Virginia's statute was constitutional because it was based on the First Amendment exception known as the true threat doctrine. According to this doctrine, the conduct restriction in the statute pursued a compelling Government interest related to the suppression of burning crosses. The First Amendment permits a state to ban true threats, which encompass those statements where the speaker communicates an intent to commit an act of unlawful violence against a group or against an individual who belongs to a legally-protected group. Unlike Brandenburg v. Ohio (1969), in this case, the Supreme Court pointed out that the speaker need not intend to carry out the threat. For the Supreme Court, a prohibition of true threats protects individuals from the fear of being a target of violence and the possibility that the threatened violence will occur. Besides, the Supreme Court regarded intimidation as a type of true threat: a speaker directs a threat to a person or group of persons with the intent to cause the victim fear of bodily harm or death. While the burning of a cross does not inevitably convey a message of intimidation, the cross-burner often intends that the recipients should fear for their lives. The Supreme Court determined that burning crosses should be considered a criminal offence in such cases.

On the other hand, the Supreme Court's majority opinion argued that the content of the *prima facie* evidence provision was unconstitutional because such provision carries the pragmatic presupposition that the burning of a cross must be automatically interpreted as an intent to intimidate the target group:

65 In Common law, *prima facie* evidence means evidence that, unless rebutted, would be sufficient to prove a particular proposition or fact.

The burning of a cross in the United States is indeed intertwined with the history of the Ku Klux Klan. Besides, it is common knowledge that the Klan has often used burning crosses as a tool of intimidation and as a threat of impending violence against African Americans. Nevertheless, the Supreme Court argued that burning crosses have also remained a distinctive political symbol of shared group identity and ideology in Klan gatherings. Although the act of burning a cross stands as a symbol of hate, it may have at least two possible meanings: (a) a person is infringing the law by performing an act of intimidation or (b) a person is engaged in political speech. The majority opinion stated that through the analysis of the specific contextual elements in each case, the actual intended meaning of burning crosses could be proven. Consequently, the Supreme Court ruled that the *prima facie* evidence provision was unconstitutional because it blurs the distinction between constitutionally banned true threats and protected political symbols.

There were two interesting dissents in the case of Virginia v. Black. Whereas the former argued that the act of burning a cross should be a First Amendment exception due to its historical association with intimidation and violence against African Americans, the latter counter-argued that burning crosses, even if it is proved that they are intended to intimidate, should not be a crime because of the statute's content-based restriction.[66] In this case, the Supreme Court only struck down the Virginia statute to the extent that it considered burning crosses as *prima facie* evidence of an intent to intimidate (a true threat).

3 The United States Court of Appeals for the Armed forces

3.1 The case of United States v. Wilcox (2008)

Wilcox[67] had been convicted for conduct that was to the prejudice of good order and discipline and discredit upon the armed forces. Wilcox appealed to the United States Court of Appeals for the Armed Forces, claiming infringement upon his First Amendment rights. The issues before the Court of Appeals for the Armed Forces were two. The first issue was determining whether the evidence at

66 In US law, a content-based restriction restricts the exercise of free speech based on the subject matter or type of speech. Such a restriction is permissible only if based on a compelling state interest and is so narrowly worded that it achieves only that purpose.

67 United States Court of Appeals for the Armed Forces. United States v. Wilcox. No. 05–0159. Crim. App. No. 20000876. Decided 15 July 2008. https://www.armfor.uscourts.gov/opinions/2008Term/05-0159.pdf (accessed 19 August 2020).

trial demonstrated that Wilcox's statements on government, race and religion constituted a crime under Article 134 of the Uniform Code of Military Justice (UCMJ). The second issue asked whether Wilcox's speech was protected under the First Amendment. At the heart of the case were the profiles the applicant had posted on an internet platform. The profiles included the following statements together with a hyperlink to a website associated with white supremacy ideologue David Lane, who had been convicted for murdering a journalist:

> I'd also like to say [. . .]
>
> I am a Pro-white activist doing what I can to promote the ideals of a healthier environment [sic]. I do not base my deeds on hate but on love for my folk's women and children. Political Affiliation is none – This government is not worth supporting in any of its components. Natures [sic] and God's laws are eternal – Love your own kind and fight for your own kind. There's no "HATE" in that!
>
> Personal quote: "We must secure the existence of our people, and a future for white children" THE 14 WORDS[68] – www.14words.com (United States Court of Appeals for the Armed Forces 2008: 31).

After reviewing the case, the majority vote of the Court of Appeals for the Armed Forces held that the evidence presented at trial was insufficient to meet the element of either service-discrediting behaviour or conduct detrimental to good order and discipline under Article 134 (UCMJ). The Court of Appeals also held that Wilcox's speech, though obscene, was under the protection of the First Amendment. I will now elaborate on the legal reasoning of the Court of Appeals. The majority vote concluded that the evidence presented at trial was insufficient to show a direct connection between Wilcox's statements and the act of directly and explicitly discrediting the military mission or environment. Besides, the reach of Wilcox's statements on the internet platform was considered too small to have an effective impact of service-discrediting on an American audience, whether troops or the general public, who may not even understand the racist tenor of the statements.

Second, it was argued that the First Amendment protects the right to freedom of expression even if it involves expressing ideas that the vast majority of society finds offensive or distasteful. Wilcox's statements, though obscene, were not found to meet any of the standards for classifying them as unprotected

---

68 "THE 14 WORDS" refers to the white supremacist slogan coined by David Lane, a member of the white supremacist terrorist group known as The Order. The fourteen words are: "We must secure the existence of our people and a future for white children". This slogan reflects the primary white supremacist worldview in the late 20th and early 21st c. The slogan urges to take immediate action against non-white people to prevent the alleged extinction of the white race. https://www.adl.org/resources/hate-symbol/14 (accessed 22 July 2022).

fighting words. In other words, for the Court's majority vote, the applicant's statements did not constitute dangerous speech – that is, words aimed at creating a clear and present danger (Schenck v. the United States 1919) or speech directed to inciting or producing imminent lawless action and likely to produce such action (Brandenburg v. Ohio 1969). The majority opinion also argued that even if a lower standard pertains in the military context, this had no application to the case. Wilcox's questioned statements were neither meant an impediment to the orderly accomplishment of any military mission, nor did they present a clear and explicit danger to military values such as loyalty and discipline.

The dissenter held that, from a legal sufficiency standpoint, direct proof of service-discrediting upon the Army was not needed. It was argued that the negative effect of Wilcox's conduct on the Army's social status could be inferred from the circumstances surrounding the case. Any person reading his profiles on the internet could understand that a member of the military – a representative of the United States – criticised the government and was a white supremacist activist. In addition, the dissenter focused attention on the huge social impact any message published on the internet can have:

> Indeed, where the internet is concerned, the impact of the metaphorical back alley protest may be magnified in time and distance in a manner distinct from that taking place in an actual back road or alley. Persons from all over the world may see it, and at a time when the street protester in uniform has long ago put the placard away, the racist message on the internet lingers (United States v. Wilcox 2008: 29).

Concerning the constitutional question, the dissenter argued that the question was whether the First Amendment protects a soldier's speech if he makes racist, service-discrediting statements publicly while holding himself out as a member of the Army. The dissenter agreed with the majority vote in that the clear and present danger standard (Schenck v. the United States 1919) or the imminence standard (Brandenburg v. Ohio 1969) were unworkable in the context of a service-discrediting case involving racist speech. In his view, the application of content-based restrictions was appropriate for the case, provided that the essential needs of the armed forces and the right to freedom of expression of Wilcox as a free American were balanced. The dissenter pointed to three national interests that are at stake in the case and enable the restriction of the right to freedom of expression of a soldier who, as a state representative, must demonstrate exemplary behaviour: (a) preventing the advent and spread of hate groups within the armed forces, (b) fostering the social image of the military as a race-neutral, politics-neutral and disciplined body and (c) recruiting and sustaining high profile members to provide for the nation's security over time.

4 The European Court of Human Rights

The European Court of Human Rights (henceforth, ECHR) is an international court based in Strasbourg, France. It consists of several judges who sit individually and are entirely independent of their country of origin. The number of judges is equal to the number of states of the Council of Europe that have ratified the Convention for the Protection of Human Rights and Fundamental Freedoms (henceforth, the Convention). The Convention lays down how the ECHR functions, and contains a list of the rights and guarantees that the states have undertaken to respect. Amongst the rights to be protected by the Convention and its Protocols are (a) the right to a fair hearing in civil and criminal matters, (b) the right to respect for private and family life, (c) the right to freedom of expression and (d) the right to freedom of thought, conscience and religion. An application may be lodged with the Convention whenever a citizen thinks there has been a violation of the rights and guarantees set out in the Convention or its Protocols. The alleged violation must have been committed in one of the states bound by the Convention.

The task of the ECHR in exercising its supervisory function is not to take the place of the competent domestic courts, but rather to review under the articles of the Convention for the Protection of Human Rights and Fundamental Freedoms the decisions they have taken in line with their power of appreciation. In particular, the Court has to ensure that the national authorities, basing themselves on an acceptable assessment of the relevant facts, applied legal standards that conformed with the Convention.

In what follows, I analyse the legal reasoning of the ECHR in a short sample of relevant cases associated with hate speech.

4.1 The case of Jersild v. Denmark (1994)

The case of Jersild v. Denmark (1994)[69] originated in an application against Denmark lodged by a Danish journalist claiming infringement of his right to freedom of expression. As a result of the publication of an article in the serious press describing the racist attitudes of "the Greenjackets" in Østerbro (Copenhagen), Jersild invited three Greenjackets and a social worker employed at the local youth centre to take part in a television interview conducted by him. During the

69 The European Court of Human Rights. Case of Jersild v. Denmark. Application No. 15890/89. Judgment. Strasbourg. 23 September 1994. Official English version of the judgment.

interview, the three Greenjackets made abusive remarks about immigrants and ethnic groups in Denmark. The interview was broadcast by Danmarks Radio as part of the Sunday News Magazine and hence, had a large audience. After the Bishop of Ålborg complained to the Minister of Justice about the television interview, the case was investigated. Subsequently, the Public Prosecutor started criminal proceedings against the three youths interviewed by Jersild, charging them with incitement to racial hatred for having made the statements cited below, which have been obtained in their English version from the ECHR's judgment. In these statements, Black people and immigrants, in general, were portrayed as wild animals:

> The Northern States wanted that the niggers should be free human beings, man, they are not human beings, they are animals.
>
> Just take a picture of a gorilla, man, and then look at a nigger, it's the same body structure and everything, man, flat forehead and all kinds of things.
>
> A nigger is not a human being, it's an animal, that goes for all the other foreign workers as well, Turks, Yugoslavs and whatever they are called.
>
> It is the fact that they are "Perkere", that's what we don't like, right, and we don't like their mentality [. . .] what we don't like is when they walk around in those Zimbabwe clothes and then speak this hula-hula language in the street.
>
> It's drugs they are selling, man, half of the prison population in "Vestre" are in there because of drugs [. . .] they are the people who are serving time for dealing drugs.
>
> They are in there, all the "Perkere", because of drugs (Jersild v. Denmark. Application No. 15890/89. Judgment. Strasbourg, 23 September 1994, p. 9).

Jersild was charged with aiding and abetting the dissemination of the Greenjackets' racist views, and the same charge was brought against the head of the news section of Danmarks Radio. The three Greenjackets, Jersild and the head of the news section were convicted in the city court. Jersild and the head of the news section appealed against the city court's judgment to the High Court of Eastern Denmark. The appeal was dismissed. Subsequently, Jersild and the head of the news section appealed to the Supreme Court. Jersild claimed that the programme aimed to draw public attention to a new phenomenon in Denmark at the time: racism and xenophobia practised by socially disadvantaged youths. He explained that he had deliberately included the offensive statements in the programme not to disseminate racist opinions but to counter them through public exposure. He also pointed out that within the context of the broadcast, the offending remarks had the effect of ridiculing their authors rather than promoting their racist views.

The Supreme Court confirmed Jersild's conviction. In the justices' opinion, Jersild had failed to fulfil his duties and responsibilities as a television journalist because he had irresponsibly encouraged the youths to make racist statements and had failed to appropriately counteract such statements in the programme.

Jersild lodged an application to the ECHR claiming a violation of his right to freedom of expression under Article 10 of the Convention. For the Court, this was the first time they had been concerned with a case of dissemination of racist remarks that deny the quality of human beings to a large group of people. At the heart of the deliberation was to balance Jersild's right to impart information and the protection of the reputation of those who have to suffer racial hate speech. For the majority opinion, the reasons given in support of Jersild's conviction were insufficient to support that the interference with his right to freedom of expression obeyed a compelling state interest. It was argued that the punishment for the statements made by another person would seriously hinder the contribution of the press to the discussion of matters of public interest. In addition, it was undisputed that Jersild's purpose in compiling the broadcast in question was not to disseminate racist ideas but to open social debate on an emerging social problem. Therefore, the majority opinion attributed more weight to Jersild's right to freedom of expression than to the rights of those who have to suffer racial hate speech.

One of the dissenters pointed out that the protection of racial minorities whose human dignity has been attacked cannot outweigh the right to impart information. Another dissenter argued that Jersild had made no real attempt to challenge the racist statements he presented in his radio interview, which was necessary if their impact was to be counterbalanced, at least for the programme's audience.

4.2 The case of ES v. Austria (2019)

The case of ES v. Austria (2019)[70] originated in an application against the Republic of Austria lodged by an Austrian national for infringement of her right to freedom of expression. According to the case record, ES had given several seminars entitled "Basic Information on Islam" at the right-wing Freedom Party Education Institute. The seminars had been publicly advertised on the website of the Institute and widely disseminated. After a participant in one of the seminars, who

70 The European Court of Human Rights. Fifth section. Case of ES v. Austria. Application No. 38450/12. Judgment. Strasbourg. Final 18 March 2019. Official English version of the judgment.

turned out to be an undercover journalist, reported on the defamatory tone of the seminar's contents, a preliminary investigation was initiated against ES. The police then questioned ES concerning certain remarks she had made during the seminars, which were directed against the Prophet Muhammad. Later, the Vienna public prosecutor's office brought charges against ES for inciting religious hatred. Amongst the remarks that the court found incriminating were the ones quoted below, which have been obtained in their English version from the ECHR's judgment:

> I./ 1. One of the biggest problems we are facing today is that Muhammad is seen as the ideal man, the perfect human, the perfect Muslim. That means that the highest commandment for a male Muslim is to imitate Muhammad, to live his life. This does not happen according to our social standards and laws. Because he was a warlord, he had many women, to put it like this, and liked to do it with children. And according to our standards, he was not a perfect human. We have huge problems with that today, that Muslims get into conflict with democracy and our value system.
>
> I./ 2. The most important of all Hadith collections recognised by all legal schools: The most important is the Sahih Al-Bukhari. If a Hadith was quoted after Bukhari, one can be sure that all Muslims would recognise it. And, unfortunately, in Al-Bukhari the thing with Aisha and child sex is written.
>
> II./ I remember my sister, I have said this several times already, when [SW] made her famous statement in Graz, my sister called me and asked: "For God's sake. Did you tell [SW] that?" To which I answered: "No, it wasn't me, but you can look it up, it's not really a secret." And her: "You can't say it like that!" And me: "A 56-year-old and a six-year-old? What do you call that? Give me an example? What do we call it, if it is not paedophilia?" Her: "Well, one has to paraphrase it, say it in a more diplomatic way." My sister is symptomatic. We have heard that so many times. "Those were different times" – it wasn't okay back then, and it's not okay today. Full stop. And it is still happening today. One can never approve of something like that. They all create their own reality, because the truth is so cruel (Case of ES v. Austria. Application No. 38450/12. Judgment. Strasbourg. Final 18 March 201, p. 3).

ES was convicted of publicly disparaging an object of veneration of Muslims: the Prophet Muhammad. The domestic court found that ES's statements conveyed the meaning that Muhammad was a paedophile. The same court found that ES had intended to accuse Muhammad of having paedophilic tendencies by making those statements. Because paedophilia is socially ostracised and outlawed, ES's statements were capable of causing justified indignation amongst Muslims.

For the domestic court, ES's remarks were not statement of fact but derogatory value judgments that exceeded the permissible limits of free speech. The court concluded that the interference with ES's right to freedom of expression in a criminal conviction had been justified as such interference had been based on law and had been necessary to protect religious peace in Austria.

ES appealed, arguing that the questioned remarks were statement of fact rather than value judgments. ES claimed that her remarks were protected by the right to freedom of expression, which includes the right to express offensive, shocking or disturbing opinions. The Court of Appeal stated that the reason for ES's conviction was that she had wrongfully accused Muhammad of paedophilia. This accusation was conveyed through defamatory words and expressions, such as "child sex", "what do we call it, if it is not paedophilia", without providing any evidence. The Court of Appeal argued that although harsh criticism of religious traditions and practices is lawful, ES had exceeded the permissible limits of the right to freedom of expression by disparaging a religious doctrine.

ES lodged a claim to the ECHR alleging that her criminal conviction had given rise to a violation of her right to freedom of expression. The ECHR found that, in this case, the domestic courts had comprehensively balanced ES's right to freedom of expression (Article 10 of the Convention) with the rights of other people to have their religious feelings protected and to have religious peace preserved (Article 9 of the Convention). The ECHR considered that the domestic courts had demonstrated that ES's remarks were not phrased in a neutral manner aimed at making a public debate concerning child marriages, but contained elements of incitement to religious intolerance and were capable of stirring up prejudice and ultimately putting religious peace at risk. The ECHR also accepted that the domestic courts had put forward sufficient and relevant reasons to support the interference with ES's right to freedom of expression under Article 10 of the Convention. Such interference, in effect, corresponded to a pressing social need and was proportionate to the legitimate aim pursued. It was finally concluded that there had been no violation of Article 10 of the Convention.

4.3 The case of Fáber v. Hungary (2012)

According to the case record of Fáber v. Hungary (2012),[71] while the Hungarian Socialist Party (MSZP) was holding a demonstration in Budapest to protest against racism and hatred, members of Jobbik, a legally registered right-wing political party, assembled in an adjacent area to express their disagreement. Fáber, who was silently holding a so-called Árpád-striped flag in the company of other people, was observed by the police. Fáber was at the steps leading to

71 The European Court of Human Rights. Second section. Case of Fáber v. Hungary. Application No. 40721/08. Judgment. Strasbourg. Final 24 October 2012.

the Danube embankment, the location where Jews had been exterminated in large numbers during the Arrow Cross regime.[72] Fáber's position was close to the MSZP demonstration and a few meters away from the square where the Jobbik demonstration was being held. The police officers supervising the scene asked Fáber to remove the flag or leave. Fáber argued that the flag[73] was a historical symbol and that no law forbade its display. The result was that Fáber was arrested by the police and subsequently convicted of the regulatory offence of disobeying police instructions. On appeal, the domestic court upheld Fáber's conviction. The court found the display of the flag offensive in the circumstances and determined that Fáber's conduct had been provocative and likely to result in public disorder in the context of the ongoing MSZP demonstration. Therefore, the court determined that Fáber's right to freedom of expression had exceeded the permissible limits, causing prejudice to public order.

Fáber filed a claim to the ECHR because, in his view, the prosecution conducted against him amounted to an unjustified interference with his rights under Articles 10 (freedom of expression) and 11 (freedom of assembly) of the Convention. In this case, Fáber's rights to freedom of expression and freedom of peaceful assembly had to be balanced with the MSZP demonstrators' right to protection against public disorder. Whereas the ECHR considered that the interference pursued the legitimate aims of maintaining public order and protecting the rights of other people, it also noted that in this context, it had not been argued that the presence of the Árpád-striped banner had resulted in a true threat or clear and present danger of violence. In other words, the display of the flag was provocative, but the domestic courts had not been able to demonstrate that it had disrupted the demonstration. Neither Fáber's conduct nor that of the other people present had been proven to be threatening or abusive, and it was only the holding of the flag that had been considered provocative. Given Fáber's passive conduct, the distance from the MSZP demonstration and the absence of any demonstrated risk of insecurity or disturbance, the ECHR determined that it could not be held that the reasons given by the national authorities to justify the interference complained of were relevant and sufficient.

---

72 The Arrow Cross Party was an extreme right-wing nationalist party, who actively supported the Nazis, and participated in atrocities against Jews, during their occupation of Hungary from 1939–1945.

73 In the US, the display of the Confederate battle flag also invites controversy. Whereas supporters associate this flag with pride in Southern heritage and historical commemoration of the American Civil War (1861–1865), opponents associate it with glorification of the American Civil War, racism, slavery, white supremacy and intimidation of African Americans.

Furthermore, the ECHR analysed whether the display of the flag in question constituted an unlawful act. Assuming that the Árpád-striped flag is a polysemic banner – it can be regarded as a historical symbol or as a symbol reminiscent of the political ideology of the Arrow Cross regime – the ECHR explained that in interpreting the meaning of an expression or symbol that has multiple meanings, the context in which the expression was uttered or the symbol displayed plays an important role. In the case under discussion, the demonstration organised by the MSZP was located at a site laden with the abhorrent memory of the extermination of Jews and was intended to combat racism and intolerance. Even assuming that some demonstrators may have considered the flag fascist, for the ECHR its mere display could not be interpreted as a true threat capable of causing public disorder.

After examining the case, the ECHR decided that the interference with Fáber's rights lacked justification and that the respondent state had breached Fáber's freedom of expression. The key foci of the legal reasoning were (1) the lawful form and political nature of the expression (the applicant was displaying a lawful flag), (2) the lack of imminent danger resulting from the expression, (3) the excessive character of the police's action and (4) the potential harshness of the sanction.

4.4 The case of A. v. The United Kingdom (2003)

On 17 July 1996, the Member of Parliament (henceforth, MP)[74] for the Bristol North-West constituency initiated a debate on municipal housing policy in the House of Commons. During his speech, the MP in question referred to a young Black woman (henceforth, A) releasing her name and home address to the public. The MP referred to A and her family as "neighbours from hell" and reported on their anti-social behaviour in the following derogatory terms:

> However, it is the conduct of [the applicant] and her circle which gives most cause for concern. Its impact on their immediate neighbours extends to perhaps a dozen houses on either side. Since the matter was first drawn to my attention in 1994, I have received reports of threats against other children; of fighting in the house, the garden and the street outside; of people coming and going 24 hours a day – in particular, a series of men late at night; of rubbish and stolen cars dumped nearby; of glass strewn in the road in the presence of [the applicant] and regular visitors; of alleged drug activity; and of all the other common regular annoyances to neighbours that are associated with a house of this type (Case of A. v. The United Kingdom. Application No. 35373/97. Judgment. Strasbourg. Final 17 March 2003, p. 4).

74 The European Court of Human Rights. Second section. Case of A. v. The United Kingdom. Application No. 35373/97. Judgment. Strasbourg. Final 17 March 2003.

The MP issued a press release of his speech to the local and national press. As a result, excerpts of the MP's inflammatory speech and photographs of A were broadly disseminated. Subsequently, A received hate mail addressed to her. One of the letters expressed racial hatred in these terms: "be in houses with your kind, not amongst decent owners" (Case of A. v. The United Kingdom. Application No. 35373/97. Judgment. Strasbourg. Final 17 March 2003, p. 4–5). Another letter threatened acts of violence against her and her family:

> You silly black bitch, I am just writing to let you know that if you do not stop your black nigger wogs nuisance, I will personally sort you and your smelly jungle bunny kids out. (Case of A. v. The United Kingdom. Application No. 35373/97. Judgment. Strasbourg. Final 17 March 2003, p. 5).

A was also stopped in the street, spat on, and demonised by strangers calling her as "the neighbour from hell." Later, a report was prepared by the Solon Housing Association, which is in charge of monitoring racial harassment and attacks. According to the report, A and her family, whom the association had already moved to another house in the past because of racial abuse, were put in a dangerous situation because her name and address had been released to the public. The family was urgently re-housed again, and the children were moved to another school too. A wrote through her solicitor to the MP outlining her complaint. She denied the truth of the majority of the MP's remarks about her family. In addition, she stated that the MP had never attempted to verify the accuracy of his statements. The MP submitted the letter to the Office of the Parliamentary Speaker, who replied that the MP's remarks were protected by *absolute parliamentary privilege* (cf. Brown & Sinclair 2020):

> That the Freedom of Speech and Debates or Proceedings in Parliament ought not to be impeached or questioned in any Court or Place out of Parliament (the Bill of Rights 1688).[75]

Parliamentary immunity is then meant to serve a compelling state interest that is so important as to justify the denial of access to a court to seek redress. Finally, A lodged a complaint to the ECHR, arguing that the MP's absolute parliamentary privilege was disproportionate because it had violated her rights under Article 6

75 The Bill of Rights 1688. https://www.legislation.gov.uk/aep/WillandMarSess2/1/2 (accessed 28 August 2020).

(Right to a fair trial) and Article 8 (Right to respect for private and family life) of the Convention.[76]

The underlying issue for the ECHR in this case was whether an MP had immunity to defame and incite racial hatred against a national citizen. The ECHR confirmed that the MP's parliamentary statements, and the subsequent news reports of them, were each protected by a form of privilege. Whereas the MP enjoyed "absolute parliamentary privilege", press coverage is generally protected by a form of "qualified privilege". This type of privilege is lost only if the publisher has acted maliciously – that is, for improper motives or with reckless contempt for the truth. The ECHR agreed with A's complaint that the comments made about her in the MP's parliamentary speech and news reports had invaded her privacy, were defamatory, and had brought dramatic consequences for her family. The ECHR argued that these factors could not modify the conclusion regarding the proportionality of the parliamentary immunity because the creation of exceptions to that immunity would challenge the protection of parliamentary speech. Consequently, the majority opinion concluded that there had been no violation of A's rights under Articles 6 and 8 of the Convention. Regarding the MP's unprivileged press release, the majority opinion argued that A could have exercised her right to access a court of proceedings in defamation or breach of confidence.

For the dissenting opinion, absolute parliamentary immunity is a disproportionate restriction of the right to access a court. In this case, both the majority opinion and the dissenting opinion opened the debate on the need to find an optimal balance between the right to freedom of expression (Article 10 of the Convention) and the right to respect for private and family life (Article 8 of the Convention). The majority opinion claimed that the new concern of modern Parliaments should be to protect the freedom of expression of their members and reconcile that freedom with other individual rights. The dissenting opinion attracted attention to the potential danger that absolute freedom of speech poses to democratic societies:

> If the freedom of speech were to be absolute under any circumstances, it would not be difficult to imagine possible abuses that could amount to a licence to defame (Case of A. v. The United Kingdom. Application No. 35373/97. Judgment. Strasbourg. Final 17 March 2003, p. 33).

76 The European Convention on Human Rights. Convention for the Protection of Human Rights and Fundamental Freedoms. Rome 4. XI. 1950. https://www.echr.coe.int/Documents/Convention_ENG.pdf (accessed 28 August 2020).

As a final point, it is also important to note that the dissenting opinion emphasised that restricting irresponsible speeches in Parliament is also a way to protect individual rights and democratic values (cf. Brown & Sinclair 2020).

5 Conclusions

Finding a balance between the right to freedom of expression and other fundamental rights is a key focus of the legal reasoning in both US and European jurisdictions. Since the 1940s, the United States Supreme Court has consistently decided that the right to freedom of expression under the First Amendment to the Constitution of the United States should be protected as a hallmark of American liberty. The Supreme Court has even established a highly protective free speech standard – the imminence standard – which even extends the protective mantel of free speech to the expression of distasteful, offensive or hateful statements, provided that such statements are not able to cause imminent lawless action. Some legal scholars claim that time has come to loosen the First Amendment in light of landmark legal decisions such as those discussed in this chapter. Several books in recent years have advocated a more European attitude towards the regulation of hate speech, because current practices in the US rob the target groups of their dignity and equal protection under the law (Delgado & Stefancic 2018; Tsesis 2009).

On the other hand, in the ECHR's legal reasoning, one can observe the effort to establish a reasonable balance between the right to freedom of expression and other fundamental freedoms. It is also significant that in Europe, freedom of expression does not have overprotection. Although Article 10 of the Convention protects freedom of expression of distasteful, offensive, shocking or disruptive messages as a sign of promoting tolerance and open-mindedness in democratic societies, the ECHR is likely to confirm the interference with the right to freedom of expression when the specific circumstances unmistakably indicate that the exercise of freedom of expression has surpassed the lawful limits, as it was the case of the domestic court decision in ES v. Austria (2019) concerning the defamation of the Prophet Muhammad. In other cases, although the ECHR confirmed infringement of Article 10 (freedom of expression), the legal decision laid out some thought-provoking ideas inviting debate and change. For example, in the case of Jersild v. Denmark (1994), the protection of human dignity was tabled. In the case of A. v. The United Kingdom (2003), parliamentary immunity was brought into question and it was suggested that irresponsible speeches in Parliament should be restricted as a way to protect freedom of expression and democratic values from abusive behaviour leading to anarchy.

Furthermore, it was shown that in both the United States and European Union jurisdictions, when courts interpret hateful statements or expressions, they analyse the pragmatic meaning speech has in the context in which it is produced. Amongst the key linguistic foci courts use for the interpretation of hate speech, one can find constant reference to the context framing the statements or expressions – that is, the circumstances that form the setting for an event, statement or idea, and in terms of which the statement can be fully understood and its meaning interpreted adequately. Another key linguistic focus for the courts to deliberate upon is the speaker's communicative intention (the illocutionary act). The intended and likely effects of the hateful act on the recipients (the perlocutionary act) are also a key linguistic focus.

The key linguistic foci referred to above are, in effect, essential categories of the linguistic discipline of pragmatics. However, the courts hardly ever request technical assistance from a linguist, as they rely on their intuitive knowledge of language.

In the second part of this book, the author will revisit some of the landmark cases studied in this chapter. In doing so, various linguistic perspectives will be adopted, each one throwing light on a different side of the multi-faceted nature of hate speech. Our purpose is to target the language clues various linguistic theories can provide that make hate speech actionable.

4 Critical discourse analysis

1 Introduction

In the early 1900s, following a symposium in Amsterdam, critical discourse analysis[77] (henceforth, CDA) emerged as a new paradigm in linguistic research. At the core of the new paradigm was a critical perspective whose gist Wodak summarised in an interview for the *FQS Forum: Qualitative Social Research* as follows:

> not taking things for granted, opening up complexity, challenging reductionism, dogmatism and dichotomies, being self-reflective in my research, and through these processes, making opaque structures of power relations and ideologies manifest. [. . .] Proposing alternatives is also part of being critical (Kendall 2007: Art. 29).

CDA is then a type of discourse analysis that goes beyond mere categorisation and description, as it tries to explain how ideologies and power relations manifest themselves in surface discourse structures. CDA encompasses several distinct approaches. Three of them are represented by Fairclough's dialectal-relational approach (2005; 2010 [1995]), van Dijk's socio-cognitive approach (1995a; 1995b; 2005; 2006b; 2015) and Wodak's discourse-historical approach (Wodak & Chilton 2005; Wodak 2011; Wodak & Reisigl 2015; Wodak & Meyer 2016 [2001]). Although the approaches mentioned are different in terms of the methodology each employed, they share the same basic principles of CDA:

a) Discourse is seen as social practice.
b) Research focuses on unveiling ideologies of discrimination and revealing structures of power and social dominance over vulnerable groups through the analysis of surface discourse structures.
c) Analysing how ideologies and power are (re)produced by discourse structures requires an interdisciplinary or even transdisciplinary approach (van Leeuwen 2005: 3–18), combining concepts and theories in history, sociology, political science and discourse analysis.
d) Because of its inherent complexity, CDA is typically divided into three levels. The macro and meso levels of analysis help the analyst investigate the bottom line of hate speech as a social phenomenon. The meso level also serves to bridge the gap between the macro and micro levels, to demonstrate how ideology (Philips 2015; Leezenberg 2017), power, dominance and

77 For an introduction to the practice of critical discourse analysis, see Bloor and Bloor (2013 [2007]).

https://doi.org/10.1515/9783110672619-004

inequality are (re)produced, propagated, legitimised, or challenged by surface discourse structures.

e) The ultimate aim of CDA is to invite reflection on power abuse and dominance, and thereby contribute to the promotion of social change and equality.

Typically, it is difficult for any type of interdisciplinary research to satisfy the scientific community's competing demands. In this respect, CDA is no exception. It has received sharp criticism from both social scientists and linguists. Whereas for historians and sociologists, CDA is too linguistic, for discourse analysts and stylists, the linguistic analysis performed by critical discourse analysts is not thorough enough, due to the extensive, though necessary, theorisation and contextualisation. In his article *What is critical discourse analysis, and why are people saying such terrible things about it?*, Toolan pointed to what, in his opinion, CDA's major weaknesses are that it:

> [. . .] needs to be more critical and more demanding of the text linguistics it uses, it must strive for greater thoroughness and strength of evidence in its presentation and argumentation, and it must not shrink from prescribing correction or reform of particular hegemonizing discourses (Toolan 1997: 101).

Wodak herself recognised that the biggest challenge CDA faces "is to implement careful and detailed linguistic analysis while also venturing into the domains of macro social theory" (Kendall 2007: Art. 29). In this vein, Machin and Mayr (2012: 208) explained that the main criticisms towards CDA have focused on interrelated issues: (a) CDA is not the only critical approach, (b) CDA is not analysis but instead an exercise in interpretation, (c) CDA does not have a methodology of its own, because the normal procedure involves integrating existing linguistic theories and methods with a critical standpoint, (d) CDA tends to ignore real discourse recipients, (e) CDA does not pay enough attention to text production, (f) CDA is too qualitative and subjective and (g) CDA is too ambitious in its quest for social change. Apart from their full discussion on criticisms against CDA, Machin and Mayr (2012) also suggested ways of extending, enriching and making CDA analysis more rigorous, by such means as adding ethnographic and corpus-based approaches to the CDA toolbox (cf. Törnberg & Tönberg 2016).

Despite criticism, it is beyond doubt that CDA has opened up the path for analysing how surface discourse structures are deployed in the (re)production of social dominance and inequality. These two social issues, social dominance and inequality, bring together CDA and hate speech in an obvious and relevant way. Hate speech is an ideal object of study for CDA, because the social phenomenon is condoned through ideologies of discrimination – e.g. racism and sexism, sustained by social practices of discrimination, and perpetuated by discourse practices of discrimination and local interactions (cf. Lutgen-Sandvik & Tracy 2012).

The present author writes this chapter with a practical agenda in mind. First, I will briefly review some central theories in CDA that I deem useful to improve our understanding of hate speech. By way of illustration, I will revisit a landmark case associated with hate speech in the jurisdiction of the United States, Brandenburg v. Ohio (1969), whose legal reasoning was discussed in Chapter 3. Although, due to space limitations I will not be able to go into all the details of the case, I hope that I will be able to demonstrate the benefits of a multilevel analysis from a CDA perspective. I claim that a CDA approach may not only help to unveil how social and discourse practices reproduce racism in Brandenburg v. Ohio (1969) but also disclose how racist speech can sometimes be invisible to the eyes of the law: At the peak of the so-called civil rights era, in 1969, Brandenburg could deliver a racist speech with some impunity. Although indicted, the extreme racist content of the Klan leader's speech was not mentioned in the indictment; he was accused of making an "inflammatory" speech, not a hateful nor a racist one.

2 Central theories in CDA

2.1 The theory of social representations

van Dijk's approach represents the socio-cognitive side of CDA (van Dijk 1981; 1995c). At the heart of this approach is the theory of social representations. This theory inspired van Dijk's *context models* (1997: 189–226) or mental representations of the structures of the communicative situation that are relevant to the discourse participants. The theory of social representations hypothesises that social actors rely on collective frames of perceptions, called social representations (cf. Wodak & Meyer 2016 [2001]: 25; Delgado & Stefancic 2018: 121). Social representations are, according to Moscovici, "systems of preconceptions, images, and values, which have their cultural meaning and persist independently of individual experiences" (Moscovici 1982: 122). Therefore, social representations comprise a bulk of concepts, opinions, attitudes and evaluations that individuals share with those belonging to the same social group. At the heart of social representations are the sociocultural knowledge and social attitudes – e.g. emotions, opinions and evaluative beliefs – shared by a specific social group (Wodak & Meyer 2016 [2001]: 26). Delgado and Stefancic also pointed out that "the shared assumptions, beliefs, and attitudes of the group members make communication between people of the same interpretive community possible" (Delgado & Stefancic 2018: 129). What is relevant about social representations for our research purposes is that they are installed in a subject's mind in such a way that they restrict the interpretation of new information. As a result, subjects will typically look for information supporting their views and disregard information that might challenge their views (Chapter 8 is devoted to Relevance theory and the pragmatic interpretation of hate messages).

2.2 The theory of ideology

Socially shared knowledge and attitudes are organised through abstract mental systems that van Dijk called *ideologies* (van Dijk 1995a: 18). These are typically (re)produced in both verbal and non-verbal discourses – e.g. pictures, posters, comics and films, and their (re)production is embedded in organisational and institutional contexts, such as mass media, social networks, parliamentary debates and court trials. van Dijk articulated the theory of ideology within a conceptual triangle that connects society, discourse and social cognition. In this approach, ideologies essentially work as:

> the interface between the cognitive representations and processes underlying discourse and action, on the one hand, and the societal positions and interests of social groups, on the other hand (van Dijk 1995a: 18).

Ideologies, which are acquired by the members of a specific social group through long-term and complex socialisation and information processing developments, are assumed, in van Dijk's view, "to control through the minds of the members, the social reproduction of the group" (van Dijk 1995a: 18). More specifically, the control function of ideologies can be divided into two specific subfunctions that van Dijk called: (a) social functions and (b) cognitive functions (van Dijk 1995a: 19). Amongst the social functions of ideologies are, for instance, organising membership admission, coordinating the group's goals and social actions and protecting the group's interests and social privileges.

Apart from social functions, ideologies have cognitive functions such as organising, monitoring and controlling the sociocultural knowledge and attitudes of the members of a specific group (van Dijk 1995a: 19). Besides, ideologies can control how individual members act, speak, write, or even understand the social practices of others (van Dijk 1995a: 20). van Dijk referred to these personal cognitions or mental representations of events, actions, or situations people are engaged in or read about as context models, which I have already referred to in section 2.1.

To explain how ideologies are (re)produced and enacted by discourse structures, van Dijk (1995a: 24–33) proposed a multilevel method based on the conceptual triangle of society, cognition and discourse. The analysis comprises five levels: (a) social, (b) cognitive, (c) social cognition, (d) personal cognition – general (context-free) and particular (context-bound) – and (e) discourse analysis – surface structures such as syntax, lexicon, local semantics, global semantics, argumentation, rhetoric, pragmatics and dialogical interaction.

2.3 The theory of power as control

From a socio-cognitive approach, van Dijk (1996; 2015) theorised *power* as a dominant group's control over a minority group. Racism, xenophobia, sexism and islamophobia are typical examples of social dominance – or hegemony – resulting in social inequality. The power of dominant groups may be enacted in various ways. For example, social power may be integrated into rules, norms and laws in the interest of dominant groups and against the interests of dominated groups. Power can also restrict access to education, jobs, public discourse and communication forms. Besides, power can control social context, discourse structures and people's minds through persuasion and even manipulation (van Dijk 2006a). CDA focuses on power abuse understood as a norm-violation that hurts the target groups, given ethical standards such as laws and human rights principles.

3 Case study: Brandenburg v. Ohio (1969)

As stated at the outset, this chapter attempts to demonstrate how CDA can improve the understanding of cases associated with hate speech. With this purpose in mind, it will be illustrative to reconsider the case of Brandenburg v. Ohio (1969),[78] discussed in Chapter 3, from a CDA perspective.

To recap, these are the facts of the case: At the request of Clarence Brandenburg, a Ku Klux Klan leader, a rally in Hamilton County was broadcast. The footage showed people wearing Klan attire, burning a cross, and included fragments of the protest speech given by Brandenburg. These criticised the President of the US at the time, Lyndon B. Johnson (1963–1969), the Congress and the Supreme

78 US Supreme Court. Brandenburg v. Ohio, 395 US 4444 (1969). Brandenburg v. Ohio. No. 492. Argued February 27, 1969. Decided June 9, 1969. https://supreme.justia.com/cases/federal/us/395/444/ (accessed 18 August 2020).

Court for colluding with African Americans and Jews against white Americans, threatening lawless action against the federal government, and inciting hatred and violence against the target groups. The first instance court in Ohio charged Brandenburg with advocating violence or unlawful methods of terrorism as a means of accomplishing political reform under the Ohio Criminal Syndicalism statute (Ohio Rev. Code Ann. § 2923.13).[79] As a result, Brandenburg was fined $1,000 and sentenced to one to ten years. His conviction was affirmed by the Supreme Court of Ohio but finally overturned by the United States Supreme Court whose primary holding was that

> [. . .] the constitutional guarantees of free speech and free press do not permit a state to forbid or proscribe advocacy of the use of force or law violation except when such advocacy is directed to inciting or producing imminent lawless action and is likely to incite or produce such action (Clarence Brandenburg v. State of Ohio, section 7).[80]

As shown in the above quote, the Supreme Court ruled that the government cannot forbid violent speech unless it is both "directed to inciting or producing imminent lawless action and is likely to incite or produce such action." The conditions that must be met to impose criminal liability for speech that incites others to illegal actions are very narrow: imminent harm, a likelihood that the incited illegal action will occur and an intent by the speaker to cause imminent illegal actions. The proposed imminence standard superseded the clear and present danger standard articulated by Justice Holmes in Schenck v. US (1919).[81] Since then, the imminence standard has remained the principal standard in the constitutional law of the United States

For the courts of justice that trialled Brandenburg v. Ohio (1969), the main issue was whether the First Amendment should protect speech that supports law-breaking or violence. Whereas for the Ohio Court and the State Appellate Court, Brandenburg's speech was not protected by the First Amendment, for the Supreme Court it was, because, in the justices' view, Brandenburg's protest speech did not incite imminent harm, imminent illegal action was unlikely to occur, and Brandenburg's intent was not to cause imminent illegal action.[82] However, I argue that the Court overlooked what was really at the heart of the case: Should the First Amendment protect racial hate speech? This question

---

79 https://law.justia.com/codes/ohio/2006/orc/jd_292313-9c3b.html (accessed 9 August 2021).

80 https://www.law.cornell.edu/supremecourt/text/395/444 (accessed 9 August 2021).

81 US Supreme Court. Schenk v. US (1919). https://supreme.justia.com/cases/federal/us/249/47/ (accessed 9 August 2021).

82 Brandenburg's illocutionary act and desired perlocutionary effects in giving his protest speech will be analysed in Chapter 6.

was specifically relevant when the case occurred and was trialled: the civil rights era (1950s–1960s) when civil rights workers were fighting for the rights to dignity and social equality of non-white Americans, and many of them were brutally attacked or murdered by the Klan (cf. Bond, Dees & Baudouin 2011).

4 The case of Brandenburg v. Ohio (1969) under a CDA approach

For analytical purposes, the CDA approach to the case of Brandenburg v. Ohio (1969) will be divided into the three levels recommended by CDA: (a) the macro level, (b) the meso level and (c) the micro-level. In an attempt to get to the source of Brandenburg's racist speech, at the macro level, the analysis will focus on racism as an ideology of dominance and on the historical background of the Ku Klux Klan, which is now legally designated as a terrorist group. At the meso level, an effort will be made to bridge the gap between the macro-level (racism) and the micro-level (Brandenburg's protest speech) by examining how racism is propagated and legitimised through the Ku Klux Klan's discourse and their influence over social, political, economic and legislative powers. At the micro-level, the analysis will show how the Ku Klux Klan's white supremacist discourse materialises in Clarence Brandenburg's protest speech at a Ku Klux Klan rally, promoting racial prejudice and inciting hatred, hostility or violence toward non-white Americans. The analysis will focus on specific surface features at lexicosemantic (racial epithets) and syntactical (agency) levels, deferring the study of surface pragmatic features to the following chapters.

4.1 The macro level: Racism

Brandenburg v. Ohio (1969) illustrates a landmark case associated with racial hate speech.[83] Although, paradoxically, the case was never trialled as such. The Klansmen share a white supremacist ideology; they falsely believe that the white race is inherently superior to other races and that whites should have power and control over people of other races. Therefore, it is reasonable to argue that white supremacy is at the heart of hate speech because it incites

---

83 For an in-depth analysis of anti-Semitism and the politics of denial against Jews, the reader may find useful Wodak (2021 [2015]), *The politics of fear: The shameless normalisation of far-right discourse* (2nd ed.). London: Sage.

hatred, hostility or violence towards so-called minority groups. The present author concurs with Lutgen-Sandvik and Tracy (2012) that, at the macro level, socially constructed categories of race and ethnicity are historically stigmatising markers that contribute to hate speech against ethnically different Others. As shown in Chapter 1, the International Convention on the Elimination of All forms of Racial Discrimination (1965) explicitly condemns white supremacist propaganda:

> State Parties condemn all propaganda and organisations disseminating ideas or theories of superiority of one race or group of one colour or ethnic origin, or attempting to justify or promote racial hatred and discrimination in any form [. . .] (Article 4).

There are many definitions of racism as an ideology of discrimination underpinning racist behaviour. However, in my view, the most comprehensive definition is that of Stollznow, who explains racism from a Natural Semantic Metalanguage (NSM) perspective:

> *racism*
> a) many people think like this
> b) there are many kinds of people
> c) people can know what kind someone is, if they can see this someone
> d) some of these people think like this about one kind of people:
> e) these are people of one kind
> f) people of this kind are not good like other kinds of people are good
> g) people of this kind cannot do many good things like other kinds of people can
> h) people of this kind are below other kinds of people
> i) people think: it is very bad if someone thinks like this
> j) very bad things can happen to some people when other people think like this (Stollznow 2017: 303).

Stollznow dissects the inner elements of racism as an ideology of discrimination. Specifically, she identifies ten elements: Element (a) introduces a cognitive scenario to suggest a common social belief. Element (b) points to the idea of race. Element (c) states a race typology based on visibly various aspects such as skin colour, hair type, eye colour and nose shape, amongst other features. Element (d) introduces another cognitive scenario because it suggests a specific attitude towards these races. In Element (e), the Agent identifies the target as belonging to a racial group. In Element (f), the Agent forms a negative evaluation of this collective group. The negative evaluation of a collective group introduces a polarisation between Us and Them, a phenomenon that van Dijk (1996; 1997) called *Othering*. In Element (g), the Agent has formed a character generalisation of the target and introduces the idea of inequality. In Element (h), the Agent perceives this collective group (outgroup) as inferior to another group (the ingroup). Element (i) depicts racism as a socially reproached attitude and behaviour. Element (j) indicates the attitude sustaining racism, which can result in negative consequences for the target, such as derogatory language, social discrimination, segregation and violence.

van Dijk (1997) pointed to the cognitive dimension of racism because it refers to the prejudice shared by a dominant group, represented in our case by the Klan. Prejudice is based on scientifically false assumptions, such as the assumption that non-whites are not as good as whites; whites and non-whites are not equal; and non-whites are inferior to whites.[84] Interestingly enough, as soon as certain patterns of collective thinking and attitudinal behaviour have established themselves in people's minds, a process that social anthropologist Hofstede (2003 [1991]) named *collective mental programming*, people must unlearn such patterns before it will be possible for them to think and behave differently, and this unlearning process is much more difficult than learning for the first time.

As mentioned earlier, ideology performs both cognitive and social functions (van Dijk 1995a). Racism, as an ideology of discrimination, performs cognitive functions – indoctrination – and social functions – protecting and defending the ingroup's interests.

4.1.1 The Ku Klux Klan: Historical background

The Ku Klux Klan (henceforth, the KKK or the Klan) has a long, violent history in the United States that has spread to other continents under different disguises.[85] In the following, I summarise the most important historical facts.

The Klan has had three periods of significant strength in American history: (a) during the Reconstruction period (1865–1877) after the American Civil War (1861–1865), (b) during the years following World War I at the beginning of the 20th century and (c) during the 1950s and 1960s when the civil rights era was at its peak. A diachronic study of this hate group reveals a constant: The white Christian American's struggle for not losing power – social dominance and control – over the other races and religious creeds in the United States. It also reveals a circular pattern describing the process of rising, development and decline of the several Klan's waves. According to Bond, Dees and Baudouin:

---

84 Stollznow (2017: 299) also explained the cognitive constituents that are at the core of any definition of racism: (1) categorisation, (2) negative evaluation, (3) stereotyping, (4) superiority versus inferiority and (5) verbal or physical manifestation.

85 For a detail account of the history of the Ku Klux Klan, see Bond, Dees and Baudouin (2011), McAndrew (2017) and Larsen (2018).

> The Klan is strong when its leaders can capitalise on social tensions and the fears of people; as its popularity escalates and its fanaticism leads to violence, there is greater scrutiny by law enforcement, the press and government; the Klan loses whatever public acceptance it had, and disputes within the ranks finally destroy its effectiveness as a terrorist organisation (Bond, Dees & Baudouin 2011: 25).

a) The Klan's first wave

Social tensions and fears came after the American Civil War (1861–1865) during the Reconstruction period (1865–1877) when white southerners had to cope with the loss of human lives, the loss of property, debts and, in their view, loss of dignity and honour because they had been defeated in the war. The Republican governments established in the South designed policies, commonly known as the Radical Reconstruction policies, to establish political and economic equality for African Americans. The so-called Radical governments granted, in effect, African Americans civil and political equality and, at least in theory, protected them in the enjoyment of the rights they were granted. Amid these profound societal changes throughout the South, the KKK is believed to have been originally founded as a secret society by six confederate veterans led by General Nathan Bedford Forrest. Later, the small society expanded into a large society, crossing class lines. Although, at the outset, the KKK had been employed as a political instrument for white Southern resistance to the Republican government, the society was soon transformed into an "instrument of fear" (Bond, Deeds & Baudouin 2011: 4). The Klansmen carried on a subversive campaign of misinformation,[86] intimidation, terror, violence and even murder against Republican leaders, activists and South sympathisers to reverse the Radical Reconstruction policies in the South. Fear was used as an instrument of psychological control of African Americans. As Fry argues:

> The whole rationale for psychological control based on fear of the supernatural was that whites were sure that they knew black people. They were not only firmly convinced that black people were gullible and would believe anything, but they were equally sure that blacks were extremely superstitious people who had a fantastic belief in the supernatural interwoven into their life, folklore, and religion (Fry 1977, in Bond, Deeds & Baudouin 2011: 11).

---

86 According to the Klan's version of history, the Northern Radical Republicans had thrown out legitimate Southern governments, bankrupted them, and replaced white Americans with African Americans in State offices. The Klan's victimising narrative also expressed the belief that the Republicans arose mobs of black savages to attack defenceless whites (Bond, Dees & Baudouin 2011: 3).

At the heart of the KKK's violent acts was the false belief that granting social equality to non-white Americans posed a dangerous threat to white Americans' social, economic and political interests. In essence, white protestants were afraid of losing power, control and social dominance over the non-white population. Although the US Congress passed specific legislation to curb Klan terrorism, the Klan managed to have full control of social, economic, political and legislative powers. Consequently, white supremacy was soon re-established in the South. It is noteworthy that the restoration of all-white governments in the South was called redemption, a term with religious denotation from which one can infer that thanks to the Klan – the saviour – the South had been saved from evil. By the mid-1870s, Southern state legislatures began enacting laws known as the Black Codes that amounted to a re-enslavement of African Americans. For example, in the State of Louisiana, the Democratic convention resolved that:

> We hold this to be a Government of white people, made and to be perpetuated for the exclusive benefit of the White Race, and [. . .] that the people of African descent cannot be considered as citizens of the United States (Bond, Dees & Baudouin 2011: 12).

When all-white conservative governments replaced the Radical governments, most of the civil and political rights African Americans had won during the Reconstruction years were rescinded. The result was the implementation of an institutionally protected system of segregation – which went under the misleading banner of "separate but equal" – that was the law of the land for more than eighty years (Bond, Dees & Baudouin 2011: 15). History tells us that everything was separate for African Americans, but that in practice nothing was equal. Although the KKK had branches in nearly every Southern State, one of its major weaknesses was that it had neither a well-organised structure nor clear leadership. This weakness, amongst others, led to its temporary decline at the very end of the 19th century.

b) The KKK's second wave

The KKK's revival was fuelled by growing hostility to the massive flow of immigration from Europe that the United States experienced in the early 20th century and widespread fear of a communist revolution. In 1915, the new Klan was founded in Georgia by William Joseph Simmons. The creation of this new Klan, as historians point out, was inspired by the romantic view of the old South, as depicted in Thomas M. Dixon's novel *The Clansman* (1905) and David W. Griffith's film *The Birth of a Nation* released in 1915. The Klan borrowed white costumes, cross burnings, rallies, and mass parades as its symbols from the film. Unlike in the first wave of the Klan, the new society was a formal fraternal organisation with national and state structure, although there were differences in how racial discrimination surfaced in each state (Larsen 2018).

The official rhetoric of this second generation of Ku Kluxers was anti-Black and focused on the threat of the Catholic church, Jews, immigrants, Asians and communists. At its peak in the 1920s, the Klan became so powerful that it was an invisible empire (Bond, Dees & Baudouin 2011: 17–23), exerting a powerful influence over the press, economy, politics and legislature of the South, as shown in the quote below:

> In Denver, Klansmen held the offices of head of public safety, city attorney, chief of police and several judgeships, and they were behind the election of its major. The Klan helped elect the State's US senators and governor at higher levels, while Ku Kluxers held four of its top offices and one seat on its Supreme Court.[87]

The KKK managed to exert strong influence over the social, economic, political, and legal systems that collectively enabled white Americans to maintain power and social dominance over people of other races and religions in the United States. Later, during the Great Depression (1929–1939), internal divisions and the individual Klansmen's criminal behaviour reduced the organisation's popularity. Besides, Federal and State bureaus actively investigated the Klan's crimes and State and local governments passed laws against the public display of Klan insignia such as cross burnings. The Klan was attacked by the press and was the focus of political hostility. Consequently, the society was cornered and temporarily disbanded in 1944.

c) The KKK's third wave

When the United States Supreme Court abolished segregation laws – separate but equal – that had been in force since the end of the 19th century, and ordered school integration in 1954, many whites throughout the South felt threatened and became determined to preserve segregation. The social tensions and fears that arose after the Supreme Court's decision echoed Southern opposition to the Radical Reconstruction government at the end of the 19th century and provided the groundwork for a revival of the Klan lead by Eldon Edwards, an automobile plant worker. When Edwards died in 1960, Robert M. Shelton, an Alabama salesman, assumed the leadership and formed the United Klans of America.

The KKK's third wave was highly violent but was much smaller than the second, peaking at no more than 50,000 members in 1965 (Bond, Dees & Baudouin 2011: 25). Nevertheless, at the time that the International Convention on the Elimination of All Forms of Racial Discrimination (1965) and the International Covenant on Civil and Political Rights (1966) were established in the international arena,

87 McAndrew (2017). The history of the KKK in American politics. *JSTOR Daily*, January 25. https://daily.jstor.org/history-kkk-american-politics/ (accessed 4 October 2021).

the Klan was responsible for a fanatical fight against civil rights activists and integration in the South of the United States. The Klan's fight led to a wave of assaults, dynamite bombings, beatings and shootings of black and white activists. These hate crimes outraged the nation and, to a certain extent, helped win support for the civil rights cause. In 1965, Klan violence prompted President Lyndon Johnson and Georgia Congressman Charles L. Weltner to call for a Congressional probe of the KKK. Seven Klan leaders were indicted by a federal grand jury and found guilty for the first time. The Klan re-emerged in the 1970s and since then its influence has helped give rise to other hate groups, amongst them the Proud Boys, a far-right group who participated in the Capitol attack in Washington DC, on 6th January 2021.

In sum, the emergence of a hate group such as the KKK has greatly contributed to legitimising and propagating racism in the United States. As evidence of this, I have shown how white supremacy has permeated the Klan's discourse and control over the social, political, economic and legislative powers in the United States over time. The pattern observed in the evolution of each KKK wave – rise, development and decline – evidences the resilience of racial prejudice in American society.

4.2 The meso level: The Ku Klux Klan's racist discourse

According to Wodak and Reisigl, racism manifests itself discursively: "racist attitudes and beliefs are produced and promoted employing discourse, and discriminatory practices are prepared, promulgated and legitimated through discourse" (Wodak & Reisigl 2015: 576). In discussing the role that discourse plays in the enactment and reproduction of racism, Wodak and Reisigl (2015: 578–579) identified four social practices: (a) marking of natural and cultural differences between the ingroup and the outgroup, (b) social construction of cultural differences, (c) hierarchisation and negative-evaluation of the outgroup and (d) legitimisation of power differences, social exploitation and exclusion. In the following, I will consider how white supremacy permeates the Ku Klux Klan's discourse, which legitimises and propagates racial prejudice towards non-white Americans.

4.2.1 A Klan rally

Brandenburg v. Ohio (1969) must be understood within the context of the Klan's third wave that arose in the South during the civil rights era in the 1950s and 1960s. As mentioned above, Brandenburg, the Klan leader, enlisted a television reporter to broadcast a Klan rally. Like many other speech events, a Klan rally is, perhaps more evidently than others, ideologically based: it serves as a social instrument of dissemination and legitimisation of racial prejudice. In the first place, ideology seeps into the representation of the participants and actions in event models. In this case, the rally participants were Brandenburg, a Klan leader and twelve Klansmen. In the second place, ideology permeates the participants' mutual relation in a Klan rally. The relationship between the Klan leader and the Klan members is asymmetric. In Brandenburg's protest speech at the rally, the ingroup was presented as "We", "us", "the whites" and "the Caucasian race", while the targets were presented as "the (dirty) nigger" and "the Jew". In addition, a third group is mentioned in the speech, "the traitors of white people", in clear reference to President Lyndon Johnson, the Congress and the Supreme Court. The historical keys to interpreting the Klan's hostility towards US authorities can be found in (a) the abolition of segregation law – equal but separate – (b) the Congressional probe of the Klan's unlawful activities and (c) the Supreme Court's conviction of Klan leaders for their lawless actions.

As a social practice, a Klan rally serves several purposes. First, from a cognitive perspective, a Klan rally assists the leader in reinforcing and controlling the ingroup's indoctrination and, maybe, attracting new members to the association. Second, from a social perspective, a Klan rally helps the leader coordinate the organisation's goals, protect and defend the ingroup's social privileges. Third, a Klan rally is used as an instrument for instilling terror in the minds of the target groups, in this case, African Americans and Jews, who may think that their lives are under threat. It is noteworthy that the Klan is presented in Brandenburg's speech as a powerful association. The excerpt below illustrates the Klan leader's claims of power:

> This is an organiser's meeting. We have had quite a few members here today which are – we have hundreds, hundreds of members throughout the State of Ohio. I can quote from a newspaper clipping from the Columbus, Ohio, Dispatch, five weeks ago Sunday morning [. . .] We are marching on Congress July 4, four hundred thousand strong. From there we are dividing into two groups, one group to march on St. Augustine, Florida, the other group to march into Mississippi (Brandenburg v. Ohio, 395 US 4444 (1969), p. U.S. 446).

The footage of the Klan rally showed firearms and munition, supporting the hypothesis that there was a serious threat of violence. Besides, one should not forget that in Brandenburg v. Ohio (1969), the legitimisation and dissemination of white supremacist propaganda were enhanced because the Klan rally was broadcast on the local station and a national network.

4.2.2 Brandenburg v. Ohio (1969): A landmark case in US jurisprudence

Ideology can also seep into and model legal doctrine. A legal doctrine is a set of rules or a test established through precedent in the common law based on which judgments can be determined in a given case. A doctrine is said to be set forth when a judge makes a ruling where a rule or test is outlined and applied, and allows it to be equally applied to similar cases. When many judges use the same rule or test, it may become established as the *de facto* method of deciding similar cases. Brandenburg v. Ohio (1969) is considered a landmark case in US jurisprudence because the Supreme Court's decision in the case changed the established doctrine in the area – that is, the clear and present danger standard was superseded by the imminence standard, also known as the Brandenburg test, for speech that advocates violence (see Chapters 2 and 3).

The clear and present danger standard, established in Schenck v. United States (1919),[88] had been the doctrine used by the United States Supreme Court to determine under what circumstances limits can be placed on First Amendment freedoms of expression, press and assembly. The clear and present danger standard had been applied to Dennis v. United States (1951),[89] in which the Supreme Court upheld the constitutionality of the Smith Act (1940),[90] which made it a criminal offence to advocate the violent overthrow of the government or to organise or be a member of any group or society devoted to such advocacy. According to the Smith Act, Brandenburg's conviction should have been upheld by the Supreme Court because his speech encouraged listeners to take revenge on the government by using unlawful methods as a means to accomplish a political reform for the benefit of white Americans:

> but if our President, our Congress, our Supreme Court, continues to suppress the white, Caucasian race, it's possible that there might have to be some revengeance [sic] taken. We are marching on Congress July 4, four hundred thousand strong. From there we are dividing into two groups, one group to march on St. Augustine, Florida, the other group to march into Mississippi (Brandenburg v. Ohio, 395 US 4444 (1969), p. US 446).

---

88 US Supreme Court. Schenck v. United States, 249 US, 47 (1919). Schenck v. United States. No. 437, 438. Decided March 3 1919. https://supreme.justia.com/cases/federal/us/249/47/ The text says that "the printed or spoken word may not be subject to restriction or subsequent punishment unless its expression creates a clear and present danger of bringing about a substantial evil" (accessed 20 August 2020).

89 US Supreme Court. Dennis v. United States, 341 US 494 (1951). https://supreme.justia.com/cases/federal/us/341/494/ (accessed 17 October 2021).

90 Smith Act (1940) by Alec Thomson. The First Amendment Encyclopaedia. https://www.mtsu.edu/first-amendment/article/1048/smith-act-of-1940 (accessed 17 October 2021).

The Supreme Court decided *per curiam*[91] that Brandenburg's conviction be overturned. The legal decision was based on the fact that freedom of speech and press do not permit a state to forbid advocacy of the use of force or law violation except where such advocacy is directed to inciting imminent lawless action and is likely to produce such action. Justice Black[92] filed a brief concurrence[93] stating that the clear and present danger doctrine should have no place in the interpretation of the First Amendment. Justice William O. Douglas wrote a separate concurrence agreeing with Justice Black's opinion.

The conviction reversal implied that the Supreme Court had not found that it was Brandenburg's intent to cause imminent harm, and harm was unlikely to happen. One may wonder which methods the Supreme Court justices employed to reach this conclusion in the case. Bear in mind that when Brandenburg gave his protest speech at the rally in 1964, he did so amid the Klan's campaign of terror against the civil rights movement: dynamite bombings, arson of black churches and killings of black people and civil rights activists in the United States. As reported in Bond, Dees and Baudoin (2011: 25), 1964 was, in fact, the same year that three civil rights workers were killed in Philadelphia, Mississippi, a Black educator was shot as he was returning to his home in Washington after summer military duty in Georgia, and a reverend was beaten during voting rights protests in Selma, Alabama. In the context of such ongoing terror, it is reasonable to argue that Brandenburg's advocacy of violent acts was not abstract but likely to represent a clear and present danger for the federal government and cause imminent harm to the target groups (cf. Bond, Dees & Baudouin 2011: 28–34). Figures 4.1 and 4.2 below show a couple of illustrative examples of hate crimes instigated by the Klan as a response to the abolition of racial segregation laws and the granting of civil rights to non-white Americans. Figure 4.1 depicts the massacre after a Baptist church was bombed in Birmingham, Alabama, in 1963. Figure 4.2 shows a missing person poster created by the FBI in 1964. The FBI investigation concluded that Klan members had murdered the three civil rights activists.

91 A *per curiam* decision is a court opinion issued in the name of the court rather than specific judges.

92 One of the justices, Hugo L. Black, was believed to have been associated with the Klan since the beginning of his career (Leuchtenburg 1973: 1–31).

93 In Law, a concurrence is a judge's or justice's separate opinion that differs in reasoning but agrees in the decision of the court.

Birmingham Post-Herald

HOME EDITION

PRICE FIVE CENTS

Alabama's 'Good Morning' Newspaper

VOL. 93—NO. 163 ★ ★ BIRMINGHAM, MONDAY, SEPTEMBER 16, 1963 22 Pages In Two Sections

BOMB BLAST KILLS 4 CHILDREN, INJURES 17 AT CHURCH HERE

ON GUARD—Birmingham police and sheriff's office personnel stood guard outside the bombed-out 16th Street Baptist Church here after victims and injured persons were pulled from the wreckage. Most of the stained glass and other windows of the church were smashed by the explosion. The above picture shows officers guarding the church, framed by the jagged edges of a smashed window.

DAMAGE — Police inspect damage to autos wrecked by dynamite blast which took lives of four Negro children at Sixteenth Street Baptist Church (background) during Sunday School yesterday.

Figure 4.1: Church bombed in Birmingham, Alabama, in 1963. Unknown author.[94]

Over the years, and especially with the advent of online hate speech, the imminence standard[95] has become a controversial doctrine (see Chapter 2); the real-life acts show that the perlocutionary effects of a rally speech, as is the case with any other type of speech, are unpredictable. Let us take a recent example from the 2020 US presidential election in the United States. In the midst of a social crisis, less serious than the one in the 1960s, President Donald Trump spoke at a Save America rally near the White House on 6th of January 2021. In his speech, Trump claimed that the 2020 presidential election had been stolen by the

94 Anonymous author. Public domain.

95 As discussed in Chapter 2, the imminence standard has been challenged by the specific communication processes of digital communication. Chapter 6 looks at this standard from the perspective of Speech act theory.

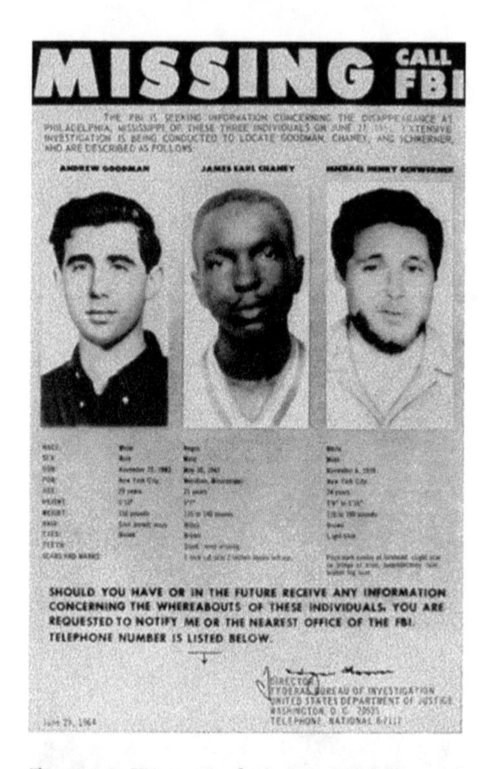

Figure 4.2: FBI poster of missing civil rights workers (1964).[96]

democrats and demanded that the Congress should reject Biden's victory. After President Trump delivered his speech,[97] a mob of his supporters assaulted the US Capitol in an attempt to overturn Donald Trump's defeat in the presidential election. The crowd broke into the building, vandalising and looting it, assaulting Capitol Police officers and reporters, and attempting to capture congressmen and congresswomen. The acts of violence ended in five fatalities. The US House subsequently impeached Trump on a single charge of incitement to insurrection. Trump's speech at a Save America rally bears a frightening resemblance with Brandenburg's questioned speech at the Klan rally and illustrates the unpredictability of the perlocutionary effects that *authoritative speech* (Philips 2004: 475) may have on its audience (see Chapter 6).

96 The poster shows the photographs of three missing civil rights workers: Andrew Goodman, James Chaney, and Michael Schwerner. Public domain.

97 Transcript of Trump Speech at Rally before US Capitol Riot. https://apnews.com/article/election-2020-joe-biden-donald-trump-capitol-siege-media-e79eb5164613d6718e9f4502eb471f27 (accessed 17 October 2021).

The United States Supreme Court's change of legal doctrine in Brandenburg v. Ohio (1969) leaves many open questions. One may wonder whether the fact that the appellant was a member of the almighty KKK might have influenced the Court's change of doctrine, taking the protection of freedom of speech to its extreme. One may also wonder whether justices can set aside their ideologies and social representations – or context models – when making a legal decision.

4.3 The micro-level: Brandenburg's protest speech

This section looks at how the Klan's racist discourse materialises in certain surface language structures in Brandenburg's protest speech. This type of authoritative speech derives its power from "the indexical connection to the leader, who is the person who is in authority" (Philips 2004: 474). Since a protest speech, in this context, invokes the moral authority of the person who delivers it, a Klan leader, it tends to be persuasive, convincing and attended to by the audience it addresses.

In Brandenburg v. Ohio (1969), the evidence the courts had to analyse was multimodal. The footage reproduced fragments of Brandenburg's speech and showed Klan insignia – cross burning, firearms and munition and a Bible. Specifically, the case record showed a scene where one could see Brandenburg wearing a red hood and robe, identifying him as a Klan leader and twelve armed hooded figures gathered around a large burning cross. Although most of the speech uttered was practically unintelligible, the following scattered utterances were presented in the case as evidence of speech inciting unlawful acts (Chapter 6 analyses the case from the perspective of Speech act theory):

> How far is the nigger going to – yeah.
> This is what we are going to do to the niggers.
> A dirty nigger.
> Send the Jews back to Israel.
> Let's give them back to the dark garden.
> Save America.
> Let's go back to constitutional betterment.
> Bury the niggers.
> We intend to do our part.
> Give us our state rights.
> Freedom for the whites.
> Nigger will have to fight for every inch he gets from now on (Brandenburg v. Ohio, 395 US 4444 (1969), p. 395 U.S. 449).

In the second scene of the first film, Brandenburg made the following remarks:

> This is an organiser's meeting. We have had quite a few members here today which are – we have hundreds, hundreds of members throughout the State of Ohio. I can quote from a newspaper clipping from the Columbus, Ohio, Dispatch, five weeks ago Sunday morning. The Klan has more members in the State of Ohio than does any other organisation. We're not a revenge organisation, but if our President, our Congress, our Supreme Court, continues to suppress the white, Caucasian race, it's possible that there might have to be some revengeance [sic.] taken. We are marching on Congress July 4, four hundred thousand strong. From there we are dividing into two groups, one group to march on St. Augustine, Florida, the other group to march into Mississippi (Brandenburg v. Ohio, 395 US 4444 (1969), p. US 446).

In the second film, Brandenburg's remarks were very similar to the ones above, except for two differences: The possibility of "revengeance" was omitted, and one remark was added: Personally, I believe the nigger should be returned to Africa, the Jew returned to Israel (Brandenburg v. Ohio, 395 US 4444 (1969), p. US 447).

In analysing Brandenburg's speech fragments, one can observe the use of an overall strategy: negative Other-presentation and positive Self-presentation. Whereas racial epithets refer to African Americans and Jews ("A dirty nigger" and "the Jews"), the Klan is depicted as the saviour of America ("Save America", "We intend to do our part"). One can also infer the legitimisation of power-differences ("Nigger will have to fight for every inch he gets from now on"), social discrimination and exclusion ("Send the Jews back to Israel", "Let's give them back to the dark garden" and "Bury the niggers") and the justification of violence ("We're not a revenge organisation, but if our President, our Congress, our Supreme Court, continues to suppress the white, Caucasian race, it's possible that there might have to be some revengeance taken."). It is noteworthy that the speech fragments of the first scene contained an imperative call for hostile and violent action against African Americans and Jews ("Bury the nigger", "Send them back to Israel", "Send them back to the dark garden"). Brandenburg strategically mitigated his call for violence in the second scene. Specifically, the disclaimer "but" and epistemic modality subtly hedged Brandenburg's commitment to his words ("possible", "there might have to be some revengeance taken"), probably as a way to avoid indictment under the clear and present danger standard. Brandenburg's stance also foregrounded the distance between the ingroup (Us) and the outgroup (Them): African Americans and Jews are portrayed as parasites and evil; the President, Congress and Supreme Court as traitors to white Americans for their support to non-white Americans, and the Klan as the saviour of America and the Caucasian race, who are depicted as victims.

Apart from these discursive strategies, Brandenburg's speech also exhibited specific topoi that are characteristic of the politics of fear (Wodak 2021 [2015]: 76). First, one can infer that the extension of civil rights to African Americans and Jews implied danger and threat to the white Americans' power, control and social domination (*Topos of danger* or *threat*). Second, the KKK is depicted as the natural saviour of America and the Caucasian race (*Topos of saviour*). Third, the US authorities should stop granting civil rights to African Americans and Jews because history teaches that this action may have violent consequences, as was the case in the first Klan wave (the Reconstruction period (1865–1877) and the second Klan wave (the years after World War I) (*Topos of history*).

Following this overview of the discursive strategies found in Brandenburg's speech, the chapter focuses on specific surface language structures at the lexicosemantic and syntactic levels of analysis that (re)produce and enact racism. (In the following chapters, the present author will return to the Brandenburg v. Ohio case and consider it from other linguistic pragmatic perspectives).

4.3.1 Surface language structures enacting racism at the lexicosemantic level

Negative lexicalisation is a major and well-known domain of ideological expression and persuasion. From Brandenburg's speech, it can be seen that racial epithets (ethnic slurs) were used to derogate the target groups.[98] Specifically, the Klan leader chose the racial epithets "nigger" and "Jew" as forms of negative lexicalisation of African Americans and the Jewish people respectively. The racial epithet "nigger(s)" was repeated eight times in the speech fragments the courts analysed, and it was even accompanied by the negative qualifier "dirty", while the epithet "Jew(s)" was used three times. The rhetorical repetition of racial epithets underpins social prejudice and hate toward the target groups.

Hom (2008) proposed the theory of combinational (semantic) externalism (CE) to explain how racial epithets get their derogatory force and what derogation with racial epithets means. According to CE, the semantic meanings of words are not completely determined by the speaker's internal mental state. The meanings of words are, instead, dependent on the external social practices of the community. Therefore, an external source determines the derogatory content of a racial epithet semantically. The plausible candidates for the external social practices grounding the meanings of racial epithets are social institutions of racism. For example, the racial epithet "nigger" is derived from and

98 Racial epithets (e.g. nigger, Jew, chink), dehumanising metaphors (e.g. parasites, bedbugs, snakes) and stereotypes (e.g. terrorists, criminals) are typical lexical features used to insult and demean the targets.

supported by racism towards African Americans. An institution of racism can be modelled as the composition of two entities: an ideology and social practices, such as subordination, alienation, dehumanisation and genocide (cf. Baider 2020: 208–210). The two entities making up racist institutions are closely related, as racists will typically justify and motivate racist social practices with their corresponding racist ideology. It is reasonable to argue that racial epithets express complex properties externally derived from racist institutions. I will now explain why these properties can be deeply derogatory and even threatening. According to Hom (2008), the meanings of racial epithets can be presented with the following complex predicate:

> *Ought to be subject to* $p^{*}1+ \ldots p^{*}n$ *because of being* $d^{*}1+ \ldots d^{*}n$ *all because of being* npc^{*}

$p^{*}1+ \ldots p^{*}n$ stands for the deontic prescriptions derived from the set of racist practices; $d^{*}1+ \ldots\ldots dn^{*}$ are the negative properties derived from the racist ideology; and npc^{*} is the semantic value of the appropriate non-pejorative correlate of the racial epithet. For instance, the racial epithet "nigger" expresses a complex, socially constructed property like:

> *ought to be . . . and ought to be, . . . because of being black, all because of being African Americans*

In Brandenburg's speech, the racial epithet "nigger" expresses a complex, socially constructed property like:

> *ought to be "buried" and ought to be "returned to Africa" because of being blank, all because being African Americans*

Similarly, the racial epithet "Jew" in the same speech expresses a complex socially constructed property like:

> *ought to be "sent back to Israel" and ought to be "sent back to the dark garden" because of being Jews, all because being the Jewish people*

As shown above, the racial epithets "nigger" and "Jew" prescribe social practices of discrimination, hostility or violence for their targets because of supposedly possessing the negative properties ascribed to their race. Racial epithets both insult and threaten their intended targets in deep and specific ways by predicting their negative properties and invoking the threat of discriminatory practice towards them. CE then explains why calling someone a racial epithet invokes an entire racist ideology and the social practices it supports.

A racist ideology can also be reproduced through local partisan semantics regarding the self-definition of the participants in the speech event. Whereas the ingroups (the Klansmen) are portrayed as saviours of America and the Caucasian race, the outgroups (African Americans and Jews) are vilified. Lexically and semantically, the targets are associated not simply with difference but rather with deviance and threat. Interestingly, the footage presented as evidence in Brandenburg v. Ohio (1969) showed a Bible, which connotes moral authority, next to firearms and ammunition. According to Metzl, "guns connote complex tensions, stereotypes, and anxieties about race" (Metzl 2019: 3). The footage also showed other images[99] popularised as potent hate symbols by the Klan: the hood and, especially, the burning cross, which can instil terror in the targeted groups. (The relevance of hate symbols will be analysed in further detail in Chapter 8). The orchestration of the linguistic and visual elements in the footage foregrounded specific ideological meanings that contextualised each other to powerfully communicate racism.

4.3.2 Surface language structures enacting racism at the syntactic level

The ideological implication of sentence structures is well known. Word order may code for underlying (or cognitive) agency (Fowler, Hodge, Kress & Trew 1979). In English, a Nominative-Accusative language, the subject of a sentence may represent a range of participants in the event, e.g. Agent, Object (also Patient), Theme, Experiencer, Goal, Beneficiary, Instrument and Locative (Duranti 2004: 460).

In Brandenburg's remarks below, one can see that syntactic topicalization emphasises the agency and responsibility of the Klan ("We", "Let's"), as the doer or instigator of the action, and the action plan to be taken ("Send", "give them back", "Save", "go back", "Bury", "Give us") to achieve their goals – that is, to stop the social advance of the outgroups (African Americans and Jews) and regain social power. The targets (African Americans and Jews) are the Object (the undergoer or patient of the action denoted by the predicate) and America and white Americans are the Beneficiary (the entity that benefits from the action denoted by the predicate).

> This is what [Action] we [Agent] are going to do [Action] to the niggers [Object]
> We [Agent] intend to do [Action] our part
> Send [Action] the Jews [Object] back to Israel
> Let's [Agent] give [Action] them [Object] back to the dark garden
> Save [Action] America [Beneficiary]
> Let's [Agent] go back [Action] to constitutional betterment

99 For a detailed study on critical discourse analysis and figures of speech, see Hart (2007), Kienpointner (2011) and Musolff (2012).

> Bury [Action] the niggers [Object]
> Give [Action] us [Beneficiary] our state rights [Object]
> Freedom [Object] for the whites [Beneficiary]

The action plan the Klan had in mind, though perverse and threatening, was presented as a set of good actions to "Save America" and demand "state rights" and "Freedom for the whites". Brandenburg's speech is fallacious because the arguments derive from faulty reasoning. In other words, no empirical evidence supports the idea that white Americans did not have civil rights or had been deprived of them by US authorities. Interestingly, the speech excerpt below presents "We" and "The Klan" as Experiencer (the entity that is aware of the action described by the predicate but which is not in control of the action) and, therefore, eluding responsibility of any violent attack.

> We [Experiencer] have had quite a few members here today which are – we [Experiencer] have hundreds, hundreds of members throughout the State of Ohio [. . .] The Klan [Experiencer] has more members in the State of Ohio than does any other organisation.

Finally, in Brandenburg's speech, social conflict is cognitively represented and enhanced by syntactic polarisation ("We" and "the niggers/Jews"), and discursively sustained and reproduced by derogating and excluding the targets from the community of American civilised people: "Bury the niggers", "Send the Jews back to Israel", "Let's give them back to the dark garden."

5 Conclusions

This chapter approached racial hate speech from a CDA perspective. The present author invited the reader to revisit a landmark case in the jurisdiction of the United States, Brandenburg v. Ohio (1969). CDA analysis was useful in exposing the Klan's racist ideology in Brandenburg's protest speech. Despite the sharp criticism CDA methods have received from various scholars and disciplines, CDA, in the my judgement, can improve the understanding of racist discourse. The description of particular lexical choices in Brandenburg v. Ohio (1969), such as the rhetorical repetition of racial epithets, revealed how the Klan leader persuasively communicated the Klan's white supremacist discourse.

The CDA approach to Brandenburg v. Ohio (1969) also invited reflection on the invisibility of racial hate speech to the eyes of the law. The United States Supreme Court reversed Brandenburg's conviction in 1969, the same year the International Convention on the Elimination of All Forms of Racial Discrimination came into force in the international arena. The fact that Brandenburg could deliver a racist speech, containing both anti-Black and anti-Semitic utterances, shows the power of a social elite and its control of American society. It was never an issue for the courts of justice that trialled the case whether Brandenburg's broadcast speech could damage the targets' rights to dignity and equality. The speech fragments analysed by the courts contained eleven instances of racial epithets – "the (dirty) nigger", "the Jew" – that are surface discourse manifestations of white supremacism. For the Supreme Court, Brandenburg's speech was under First Amendment protection, despite his words implied xenophobia, racial prejudice and resentment towards non-white Americans. I concur with Matsuda, Lawrence III and Delgado that "the failure to provide a legal response limiting hate propaganda elevates the liberty interests of racists over the liberty interests of the targets" (Matsuda, Lawrence III & Delgado 1993: 40) and perpetuates racial prejudice and social inequality.

Apart from disclosing the role language plays in the maintenance of social inequalities (Philips 2004: 495), a CDA approach also seeks answers to some relevant questions concerning hate speech regulation: Why hate speech was not – and even today is still not – unprotected speech under the constitutional law of the United States?[100] Critical race theory (CRT) is a liberal movement that emphasises the historical experiences of oppressed minorities to effect political change and eradicate racism. A specific part of their project argues that racial epithets fail to merit First Amendment freedom of expression protection and should be assimilated into fighting words and true threats. If successful, the argument would allow for government and institutional restrictions on the use of racial epithets and hence, protect minority groups from group defamation (cf. Delgado & Stefancic 2018). In this respect, the analysis of racial epithets through combinational externalism (CE) (Hom 2008) offered well-motivated semantic reasons for literally assimilating racial epithets to true threats.

CDA can help improve the understanding of how little progress hate speech regulation has made in the United States. Delgado and Stefancic argued that hate speech regulation does not pose any limitation to the First Amendment

> because it implicates the interest of minorities in not being defamed, reviled, stereotyped, insulted, [. . .] and harassed. Permitting a society to portray a relatively powerless group in this fashion can only contribute to a stigma picture or stereotype according to which its members and unworthy of full protection (Delgado & Stefancic 2018: 155).

100 The following forms of speech are unprotected under the United States constitutional law: (1) obscenity, (2) fighting words, (3) defamation, (4) child pornography, (5) perjury, (6) blackmail, (7) incitement to imminent lawless action, (8) true threats, (9) solicitation to commit crimes, (10) treason if committed verbally and (11) plagiarism of copyrighted material.

The above quote illustrates how legal practice can perpetuate the targets' stigmatisation, stereotyping and marginalisation. Minority groups carry a stigma that silently hinders their full access to education, employment and participation in public life.

A CDA approach to racial hate speech provides logical and reasonable arguments favouring a legal change. For Delgado and Stefancic, the solution to the legal problem is by no means a simple task, because judges are not free from the so-called observer's paradox:

> Judges asked to strike a balance between free speech and minority protection decide the contours of a new interpretive community. They must decide whose views count, whose speech is to be taken seriously, and whose humanity is fully respected. Can they do so fairly and open-mindedly, since most of them come from the dominant speech community? (Delgado & Stefancic 2018: 132).

5 Register and genre perspectives on hate speech

1 Introduction

Chapter 4 explained how racist discourse is encapsulated in ideologically-based linguistic surface structures, such as racial epithets, polarisation, partisan semantics and agency. This chapter elaborates on the surface linguistic structures that articulate hate speech. Specifically, the analysis in this chapter concentrates on texts or discourse segments of various dimensions in which hate speech manifests itself and on the genre or genres (Solin 2011) into which hate texts can be classified. The discussion attempts to answer two essential questions: Is there a hate register? Is there a hate speech genre?

2 Discourse, texts and genres

Reisigl and Wodak explained how the concepts of discourse, texts and genres are intertwined:

> discourse is a complex bundle of simultaneous and sequential interrelated linguistic acts that manifest themselves within and across the social fields of action as thematically interrelated semiotic, oral or written tokens, very often as *texts*, that belong to specific semiotic types, i.e. genres (Reisigl & Wodak 2001: 36).

One can learn from the above quote that hate speech, like any other type of discourse, can manifest itself through thematically interrelated texts linked to other texts over time. This type of relationship, which is commonly referred to as *intertextuality* (Kristeva 1980; Bakhtin 1981; Bazerman 2004), can be established, as Reisigl and Wodak suggested, in a variety of ways: (a) explicit reference to a topic, (b) references to the same events, (c) allusions or evocations and (d) transfer of main arguments from one text to another (Reisigl & Wodak 2001: 28).

Therefore, hateful texts may be considered parts of hate speech that express *misethnicity*, a concept Tsesis defines as “hatred towards groups because of their racial, historic, cultural, or linguistic characteristics” (Tsesis 2002: 81). Hatred is linguistically embodied in hateful expressions and speech acts. Texts make such hateful expressions and acts durable over time and, therefore, bridge two speech situations that are not usually adjacent in time and space: the speech production and the speech reception. The non-adjacency in time and space between the production and reception of “destructive messages”

https://doi.org/10.1515/9783110672619-005

(Tsesis 2002) challenges the imminence standard for adjudicating cases of criminally inflammatory speech in the American legal system (see Chapters 2 and 3). I will go back to this controversial legal standard in Chapter 6, which addresses hate speech from the perspective of Speech act theory.

According to Biber and Conrad (2009), texts can be analysed from three perspectives: (1) the *register perspective*, (2) the *genre perspective* and (3) the *style perspective*. For purposes of analysis, this chapter focuses on the former two perspectives, leaving aside the latter. The register and genre perspectives share the concept that texts can be described according to their contexts, considering the situation, the participants and the communicative purposes associated with the texts. Both perspectives differ in two essential respects. First, whereas the linguistic component of register analysis describes the linguistic features that are pervasive in the text variety, the linguistic component of genre analysis describes the conventional structures, which may only appear once in the text variety. Second, register analysis can be done on both text excerpts and complete texts, while genre analysis requires complete texts.

3 A register perspective on hateful texts

Biber and Conrad (2009) adopted a systemic functional perspective on the concept of register. In their view, the register perspective offers an analytical framework for studying texts that consists of three elements: (1) the situational context, including the physical setting, the participants and the different roles they play in the communicative situation, (2) the communicative purpose(s) and (3) the linguistic features that are pervasive in the text variety. Biber and Conrad's proposed analytical framework narrowly draws on Halliday's *context of situation*, whose purpose is to explain "how a text relates to the social processes within which it is located" (Halliday 1978: 10). Context of situation consists of three parameters known as semiotic functions: (1) *Tenor*, (2) *Field* and (3) *Mode*. Tenor concerns the participants, their roles and the type of social relationship between them. Field refers to the topic or area of external reality with which the text deals. Mode refers to the channel through which communication occurs and to text construction. In the following, I will examine each of the aforementioned semiotic functions, Tenor, Field and Mode, in detail.

3.1 The context of situation describing a hate register

A hate register can be described through the context of situation in which the text is anchored. The Tenor is diverse: The speaker often belongs to a dominant group, whose speech is driven by prejudice and intolerance towards members of a group identifiable by legally-protected characteristics, while the addressees can be of different types. On the one hand, there are the ingroups, with whom the hate speaker shares the same ideology, prejudice and intolerance. On the other hand, there are the outgroups, who are the object of hate. There may also be overhearers, bystanders and eavesdroppers. These latter people could be eventually indoctrinated and recruited into the association or organisation. The roles and types of relationships established and maintained between the speaker, the ingroups and the outgroups are complex. Characteristically, the speaker plays the role of the saviour, the ingroups represent the victims, and the outgroups signify a dangerous threat putting the ingroups' social privileges and economic interests at risk. van Dijk (1992) called this role shift *victim-perpetrator reversal*. In addition, the types of relationships that the participants establish and maintain between them may be of two kinds: (1) a relationship of solidarity between the hate-advocating speaker and the ingroups and (2) a relationship of power and social dominance between the hate-advocating speaker and the outgroups.

The Field typically contains fallacies about the outgroups' moral inferiority or non-humanity and false statements of fact harmful to their dignity and social reputation (Delgado 1982; Matsuda 1989; Waldron 2012; Brown 2017b).

The Mode can be varied. First, hateful texts can be conveyed through any communication channel: oral, written, or multimodal – that is, combining language and other means of communication, such as images, sound recordings and footage, amongst other possibilities. Second, hateful texts can be either non-interactive or interactive, or both. Third, hateful texts can be non-computer mediated or computer-mediated. Fourth, the hateful text's rhetoric is characteristically manipulative, thereby appealing to the emotions (*pathos*) of the members of the intended audience.

3.2 The set of communicative purposes describing the register of hateful texts

In theory, registers are described for their primary communicative purposes. However, defining a register's communicative purpose appears to be indefinite and sometimes subjective. I concur with Bhatia that "if one was looking for clear-cut, definite and objective criteria to define and identify communicative purposes for each genre, one would necessarily be frustrated by the complex realities of the world of discourse" (Bhatia 2014 [2004]: 130).

Marwick and Ross (2014) pointed to intent-based hate speech as one of the core elements of hate speech (see Chapter 1). I argue that the register of hateful texts cannot be described by a single communicative purpose but, instead, by a set of communicative purposes, such as to disseminate, advocate and incite hatred, hostility or violence against the members of a target group, without communicating any legitimate message (Moran 1994; Ward 1997; Benesh 2014; Marwick & Ross 2014).

3.3 The linguistic features describing the register of hateful texts

Registers can also be described through the pervasive linguistic features of the text. Critical discourse analysts (see Chapter 4) have provided useful insights into several characteristic ideological structures common in racist texts (and talk). Specifically, van Dijk (2011) pointed to the following set of *ideological structures*:[101]

a) *Polarisation*: the positive representation of the ingroup – e.g. the glorification of our country contrasts with the negative representation of the outgroup, typically depicted as outsiders and invaders.
b) *Pronouns*: racist speakers use the first-person plural form of the personal pronoun (We) – in its various grammatical forms (us, our, ours) – to refer to themselves and fellow members, while they use the third-person plural form of the personal pronoun (They) – in its various grammatical forms – (them, their, theirs) to refer to the members of a target group.
c) *Ideological square*: racist speakers combine hyperbolic emphasis when referring to the positive aspects of the ingroup along with minimisation of the ingroup's weaknesses or the outgroup's strengths.
d) *Activities*: racist speakers employ *deontic modality* to refer to the activities the ingroups do or must do to either protect the ingroup's social privileges and economic interests or to marginalise, attack or control the outgroup.
e) *Norms and values*: racist speakers refer to the social norms and values the ingroups must strive for – e.g. Freedom and Justice.
f) *Interests*: racist speakers refer to the ingroup's interests, such as their material or symbolic resources.

101 Although van Dijk considered the ideological structures common in racial hate speech, they may be equally applied to any discourse expressing hatred towards other legally-protected groups.

Some of the above ideological structures can be seen, for instance, in Brandenburg v. Ohio (1969) (see Chapter 4). Brandenburg's speech contains polarisation (the positive presentation of the ingroup v. the negative presentation of the outgroup), pronouns ("We" v. "They"), activities ("This is what we are going to do to the niggers", "Let's give them back to the dark garden", "Let's go back to constitutional betterment", "Bury the niggers"); and interests ("Give us our rights", "Freedom"). It should be pointed out that for each of these ideological structures, there may be local generic forms expressing them (van Dijk 2011). In other words, the linguistic features that are pervasive in a particular genre may not be pervasive in another. Therefore, proposing a closed catalogue or inventory of the characteristic linguistic features describing a hate register may be impractical and fruitless.

Despite the inherent difficulty in determining the characteristic linguistic features of a hate register, CDA researchers have recently started to use corpus linguistics in an attempt to identify such features. For example, Brindle (2016) combined CDA and a corpus linguistics (CL) methodology,[102] using the Sketch Engine tool, to analyse white supremacist language in a web-forum dealing with homosexuality on the avowedly white supremacist website Stormfront. In his research, Brindle analysed the lexicon employed in the construction of heterosexual white masculinities, gay men, racial minorities and other outgroups. Brindle's study revealed several high-frequency terms that are pervasive in Stormfront's homophobic texts. These terms are:

> "queer(s)", "homo(s)", "fag(s)", "faggot(s)", "faggotry", "pervert(s)", "perversion", "pedophiles", "paedophilia", "sodomite", "sodomy", "molest", "molesters", "molestation", "degenerate" and "degenerates".

One should be cautious about establishing a correlation between a word's high frequency of occurrence in a text with its status as a characteristic linguistic feature of a hate register. In this regard, Brindle (2016) hit the nail on the head when he argued that the words not present on a frequency list may just be as noteworthy as those included. In this vein, in analysing racial epithets, Technau claimed that "hate speech can be identified as the most central, albeit not the most frequent, mode of use" (Technau 2018: 25). The same author explained that other non-referential and non-pejorative uses of racial epithets may be more pervasive in other contexts that are free of hate speech – e.g. mobbing

102 Brindle (2016) applies three types of analysis to the corpus: (1) a corpus-driven approach centred on the study of frequencies, keywords, collocation and concordance analysis, (2) a detailed qualitative study of posts from the forum and the threads in which they are located and (3) a corpus-based approach which combines the corpus-linguistic and qualitative analyses.

and insulting; appropriation in counter-speech; rap, in which racial slurs, such as *nigga* are applied to the members of a target group (referential uses) by a member of the same target group (non-pejorative uses); mock impoliteness in youth language and banter – racial or homophobic terms are applied to non-members (non-referential uses) or to people for whom the speaker has an affection (non-pejorative uses); neutral mentioning in academic discourse; and incognizant uses.

I argue therefore that it might be more productive to look for the features in a text that have *collective salience* (Kecskes 2014: 184) for the ingroup than searching for the pervasive linguistic features of a text (Biber & Conrad 2009). Whereas *pervasiveness* is based on *statistical significance*, collective salience is grounded in *prominence*. Prominence does not come from the language itself. On the contrary, it is external to it, because it is the result of prior experience and conventionalisation shared with the other members of the speech community.[103] I concur with Technau (2018: 27) that the collective salience of racial epithets, such as nigger, kike, kraut and other group-based slurs - faggot and spaz, is framed by the context of situation and the communicative purpose(s) of such texts. Specifically, the context of situation must include a hate-advocating speaker communicating hateful messages that are directed at the members of a legally-protected group. In this context, some linguistic features may have collective salience because they have the power to invoke social prejudice, hostility or violence against a target group. Such power derives from the ideas, beliefs, views and attitudes the hate-advocating speaker shares with the audience of the texts. The common ground shared by the hate-advocating speaker and their audience facilitates the recognition of the linguistic features that are salient in a hate register.

Although most of the research on hate speech has focused on identifying the overt lexico-semantic features typical of a hate register, a recent publication edited by Knoblock (2022) breaks new ground in demonstrating how grammatical features such as morphology (Mattiello 2022: 34–58; Tarasova & Sánchez Fajardo 2022: 59–81), word formation (Beliaeva 2022: 177–196), pronoun use (Thál & Elmerot 2022: 97–117; Lind & Nübling 2022: 118–139; Flores Ohlson 2022: 161–176), verb mode (Bianchi 2022: 222–240) and syntactic structures are appropriated, manipulated and exploited by hate-advocating speakers to express hate speech covertly. The analysis of the grammar of hate speech marks a turn in corpus-assisted discourse studies, which typically focus on the frequency, keyness and semantic

103 See Chapter 4 for the theory of combinational (semantic) externalism (CE) (Hom 2008). Chapter 8 also looks at the role collective salience plays in speech production and speech interpretation.

prosody of words. According to Geyer, Bick and Kleene (2022: 242), whereas words and expressions are easy to detect automatically on internet-based social networks, grammatical constructions are much more difficult to identify and remove from a platform.

An additional difficulty in determining the linguistic features describing a hate register comes from their *implicitness* and *indirectness* (see Chapters 6 and 8). In examining the present limitations of corpus linguistics, Ruzaite argued that qualitative analysis can be more informative because they hate-advocating speaker ever-increasingly resorts to “creative language use” (Ruzaite 2018: 110). For example, let me comment on the use of *topoi* (conclusion rules) in argumentation (Wodak & Meyer 2016 [2001]). Topoi are often expressed implicitly but can be made explicit as conditional or causal paraphrases, such as *If X, then Y* or *Y is the result of X*. As a result, hateful messages may be conveyed implicitly, leaving almost no trace for an algorithm to detect at the surface level. At present, hate-advocating speakers are aware of social reprobation and hence in many jurisdictions they tend to convey their destructive messages implicitly and indirectly to avoid indictment and criminal prosecution.

4 A genre perspective on hateful texts

As mentioned at the outset of this chapter, the genre perspective shares with the register perspective the idea that texts can be described according to the context of situation and the set of communicative purposes associated with them. In Christie’s view, genre theory differs from register theory in terms of “the amount of emphasis it places on social purpose as a determining variable in language use” (Christie 1987: 59). In this vein, I claim that the difference between these theories goes far beyond a mere question of emphasis, because genre theory aims to provide a theory of language use. I concur with Bhatia that:

> genre theory gives a grounded or what sociologists call a *thick* description of language use rather than a surface-level description of statistically significant language features, which has been very typical of much of register analysis (Bhatia 1996: 40).

Genre theory has been traditionally applied to “the study of situated linguistic behaviour in institutionalised academic or professional settings” (Bhatia 2014 [2004]: 26), whether defined as *typification of rhetorical action* (North American orientation), as in Miller (1984), Bazerman (1994) and Berkenkotter and Huckin (1995); *staged goal-oriented social processes* (Australian orientation), as in Martin, Frances and Rothery (1987), Christie (1987) and Martin (1993), or *conventionalised communicative event* (British tradition), as in Dudley-Evans (1986), Swales

(1990) and Bhatia (1993;1997; 2014 [2004]). Despite the various orientations emerging from applications of the theory in academic and professional contexts, genre theory, as Bhatia argued, "does seem to have a common paradigm, a coherent methodology, and an overlapping concern with applications" (Bhatia 1996: 41). According to the same author, the three orientations to genre theory share five essential features (Bhatia 1996: 47–54):

1) *emphasis on conventions*: genres are based on shared communicative purposes, rhetorical conventions and regularities in structural forms;
2) *dynamism*: genres are dynamic rhetorical structures that can be exploited and manipulated according to conditions of use;
3) *propensity for innovation*: genres gradually change over time in response to new socio-cognitive needs of users;
4) *generic versatility*: genres can adapt their set of communicative purposes depending on the context of situation; and
5) *genre knowledge*: genres imply knowledge about particular processes of genre construction, dissemination and interpretation.

Genre theory has been successfully applied to analysing academic and professional types of discourse. One can presume therefore that its application to the analysis of language crimes, in this case hate speech, can contribute to widening the scope of genre theory and provide it with real-life input data from social contexts other than the academic and the professional.

4.1 Genre as typified rhetorical action

Genres may be seen as typical responses to recurring rhetorical situations. Although Bitzer does not use the term genre, his account of rhetorical situations clarifies the way genres are constructed as *typified text-types*:

> From day to day, year to year, comparable situations occur, prompting comparable responses; hence rhetorical forms are born, and a special vocabulary, grammar, and style are established. The situations recur, and because we experience situations and the rhetorical responses to them, a form of discourse is not only established but comes to have a power of its own (Bitzer 1968: 13).

In a similar vein, Bazerman (1994) defines genre as typified utterance and intention, and explains the concept in these words:

> over a period of time, individuals perceive homologies in circumstances that encourage them to see these as occasions for similar kinds of utterances. These typified utterances, often developing standardised formal features, appear as ready solutions to similar

> appearing problems. Eventually, the genres sediment into forms so expected that readers are surprised or even uncooperative if a standard perception of the situation is not met by an utterance of the expected form (Bazerman 1994: 82).

The view of genre as typified utterance and intention has stimulated research into genre construction, with special emphasis on "rhetorical conventions" and "typical textualization patterns" (Bhatia 2014 [2004]: 27–28). This emphasis has, to some extent, promoted the false assumption that genres are static linguistic artefacts, highly predictable and easily identifiable because of their conventionalised rhetorical patterns (Giltrow 2013). Nevertheless, one of the characteristics of hate speech is that it is not bound to a specific form, and hence it is highly unpredictable. Let me give some substance to this assertion by taking up the various genres in the court cases associated with hate speech we looked at in Chapter 3:

a) Terminiello v. Chicago (1949): a speech delivered by a priest in a massive auditorium.
b) Brandenburg v. Ohio (1969): a speech delivered by a Klan leader to a large group of Klansmen. The speech was even broadcast on television.
c) National Socialist Party v. Skokie (1977): a massive demonstration.
d) Jersild v. Denmark (1994): an article published in the serious press describing the racist attitudes of a racist group in Østerbro (Copenhagen) and a television interview of the leaders of such group.
e) Virginia v. Black (2003): a speech delivered by a Klan leader to a large group of Klansmen and burning crosses displayed in public places.
f) A. v. United Kingdom (2003): a political speech on municipal housing policy delivered in the House of Commons and a press release of the same political speech including photographs.
g) United States v. Wilcox (2008): profiles posted on an internet platform and a website associated with white supremacy ideologue David Lane.
h) Fáber v. Hungary (2012): a demonstration of members of Jobbik political party showing their racist challenge to an anti-racist and anti-hatred demonstration held by the Hungarian Socialist Party. The so-called Árpád-striped flag was also prominently exhibited at a location where Jews had been exterminated in large numbers during the Arrow Cross Regime.
i) ES v. Austria (2019): a series of seminars entitled *Basic Information on Islam*. The seminars were advertised on the website of the right-wing Freedom Party Education Institute.

The above court cases share, in general terms, a similar context of situation: (a) the hate-advocating speaker is a member of a dominant group, (b) the addressees are members of the ingroup, but there may also be other types of recipients – e.g. overhearers, bystanders and eavesdroppers, (c) the target is a member of a legally-protected group and (d) the contents are racist, anti-Semite or show religious intolerance. In addition, the communicative purposes are associated with disseminating, advocating or inciting hatred, hostility or violence against the members of a target group. Finally, it can be said that the lexicon, images and symbols employed by the hate-advocating speakers have collective salience for the ingroups.

Hate speech is not bound to a specific generic form. On the contrary, a variety of domain-specific genres are employed, such as protest speeches, demonstrations and parliamentary speeches (domain of politics); seminars (domain of education); press articles, press releases and TV interviews (domain of journalism). Besides, the genres employed by the hate-advocating speakers often include symbols that carry associations with a long history of hatred against the targets. For example, whereas burning crosses are hate symbols for African Americans, Swastikas are hate symbols for the Jewish people.

The genres employed in the aforementioned cases share three features: (1) they express authoritative discourse (O'Connor & Michaels 2007; Guzmán 2013), (2) they are persuasive forms of communication (van Dijk 2006a) and (3) they ensure the public dissemination of hate speech through mass media or the internet.

The above discussion shows that hate speech cannot be ascribed to a single genre identifiable by specific rhetorical conventions or textualised patterns. This difficulty in generic adscription makes the recognition and interpretation of hate speech challenging, especially for those who do not belong to the same speech community (the ingroup) and, what is more relevant, hate speech may pass unnoticed in the eyes of the law.

4.2 Genre as typified social action

Genre can be seen as social action embedded within disciplinary, professional and other institutional practices. Hence, what the participants in the speech event recognise, apart from a highly conventional rhetorical structure, is a specific *social action* (Giltrow 2013), a concept originally used by Miller in her seminal work on genre theory:

> Genre refers to a conventional category of discourse based on a large scale typification of rhetorical action; as action, it acquires meaning from situation and from the social context in which that situation arose (Miller 1984: 163).

Drawing on Miller's approach to genre as *typified social action*, Giltrow claimed that a genre should be "known not only by its formal manifestation but also by its motive" (Giltrow 2017: 48–49). The same author further argued that in "each situation, speech has cleared to a sphere of activity and has become characteristic of it" (Giltrow 2017: 49). From a wider social perspective, hate speech may, in effect, be associated with a specific sphere of negative social action, and it has become, over the years, characteristic of it, giving rise to what we know as *hate propaganda*.

5 Hate propaganda

Hate speech must be propagated to fuel hatred and hence incite hostility or violence against the members of a legally-protected group. The connection between hate speech and propaganda – a promotional genre – is clear. Hate propaganda may be categorised as a negative type of propaganda, because it promotes an ideology that incites prejudice and intolerance against the targets. Hate propaganda relates to adversarial communication, especially of a biased or misleading nature. It is designed to tap into people's deepest values, fears, hopes and dreams for the purpose of influencing their emotions, attitudes, behaviour and opinions for the hate group's benefit (cf. Chilton 2011). From a formal point of view, one of the characteristic features of hate propaganda, as is the case with any type of propaganda, is that it can operate in all genres. For this reason, hate propaganda is not bound to a particular rhetorical form. To illustrate the diverse genres that may be appropriated and exploited by hate-advocating speakers, I will examine some old and well-known genres employed in racist propaganda. Postcards were commonly used in racist propaganda against Asians in the US at the beginning of the 20th century. Figure 5.1 reproduces a postcard from 1907 depicting a demonised image of a Chinese person.

Nazi propaganda was orchestrated by Joseph Goebbels.[104] It is a well-known fact that the Nazis advocated clear messages tailored to a broad range of German people. These messages were intentionally and strategically designed to exploit and manipulate people's fear of uncertainty and instability. Jews and communists featured heavily in the Nazi propaganda as enemies of the German people (Tsesis 2002: 11–27). Figure 5.2 illustrates Nazi anti-Semitic propaganda. This two-dimensional work of art is a complex visual metaphor that dehumanises

104 In 1926, Joseph Goebbels was Gauleiter of Berlin. Later, in 1933, he was Reich Minister of Nazi propaganda.

Figure 5.1: A racist postcard: The Yellow Peril and a Chinese man by Fred C. Lounsbury (1907). Public domain.[105]

Jews and vilifies the enemies of Nazi Germany. Specifically, the visual metaphor's topic (or subject) is the Jewish people, satirically portrayed as the wandering Jew. The vehicle (or element used metaphorically) is the image of a giant worm or parasite on the earth. The worm's eyes mirror two well-known symbols representing the enemies of Nazi Germany: the hammer and sickle, which symbolise the dangers of communism, and the dollar, which stands for capitalistic greed. The meaning of

105 The postcard text reads: "He's a Yellow Peril Chink of surprising versatility. And he stole into our country with astonishing facility; he'll washee-washee shirtee for the Melican Gentility. And sit with girls in Sunday School in studious humility. Ah Sin's the heathen's name." Fred C. Lounsbury, "He's a Yellow Peril Chink . . .," *Chinese Immigration in the Late 19th Century*. http://projects.leadr.msu.edu/progressiveeraimmigration/items/show/16 (accessed July 28, 2022).

the metaphor is derived from the ground (the relationship between the topic and the vehicle): Jews are like vermin. Hence, they are a powerful threat to the world.

Figure 5.2: Nazi anti-Semitic propaganda at Yad Vashem. Faithful photographic reproduction of a two-dimensional, public domain work of art by David Shankbone.[106]

Figure 5.3 shows a poster combining images and text. The text contests Pope Pius XI's assertion that the whole of humanity is a single great catholic race, arguing instead that racial difference is inherent. The text also suggests that mixed marriages are ignominious. The superiority of the Aryan race is conveyed by a partisan contrast between the fine pictures of a young German man and woman and the ugly pictures of three Black people in stereotypical roles: the African savage, the servant and the American soldier. The selected images illustrate racial difference and endorse white supremacy.

Apart from posters and postcards, Nazi hate propaganda appropriated, exploited and manipulated other generic forms, such as cartoons, pamphlets, books, speeches, radio and television broadcasts, legal norms and rules and literature, amongst others. Figure 5.4 displays some examples of Nazi racist literature.

---

106 https://upload.wikimedia.org/wikipedia/commons/f/fd/Nazi_Anti-Semitic_Propaganda_by_David_Shankbone.jpg

Figure 5.3: Racist propaganda of the Nazi regime. German magazine from the early 1940s. Unknown author. Public domain.[107]

In sum, it has been shown that one of the characteristic features of hate propaganda is that it is not bound to a particular rhetorical form, but instead appropriates, exploits and manipulates other generic forms. None of the generic forms employed in hate propaganda was originally created to disseminate, advocate or incite hatred, hostility or violence against the members of a target group. In the next section, I will try to explain this phenomenon by looking at the concept of *genre-bending*.

107 https://upload.wikimedia.org/wikipedia/commons/f/fd/German_nazi_rasist_propaganda.jpg

Figure 5.4: Examples of Nazi racist literature. Photograph by Thomas Quine.[108]

5.1 Genre-bending

Genre-bending occurs by appropriating generic resources from a specific genre to construct more dynamic or innovative generic forms (Bhatia 2014 [2004]: 99–127). In the specific case of hate propaganda, it was shown that hate propaganda does not fit neatly into a single and pure generic form. On the contrary, hate propaganda colonises other genres, a phenomenon that Bhatia called "invasion of territorial integrity" (Bhatia 2014 [2004]: 100). Hate propaganda has, in effect, invaded academic, corporate, political, journalistic and digital genres, displaying non-conventional uses of generic resources. Genre colonisation is possible because far from the general belief that genres are static, they are, in effect, characterised by *dynamism, innovation* and *versatility* (Berkenkotter & Huckin 1995). Individuals or groups can manipulate generic conventions to communicate their private intentions within a genre's socially recognised communicative purpose. It can

108 Auf gut deutsch Wochenschrift für Ordnung u. Recht 1. Jahrgang 1919, Wilhelm Schmidt: Rasse und Volk, Hans F. K. Günther: Rassenkunde Europas. Munich City Museum (Münchner Stadtmuseum), Germany 2014.

then be argued that hate propaganda producers have knowledge about particular production processes, dissemination and consumption of genres (Fairclough 1992), because they select the genres that can help them best accomplish their malicious communicative purposes.

Genre colonisation may range from a relatively subtle appropriation of lexicogrammatical and discoursal resources to hybridisation, mixing and embedding of genres. Genre colonisation challenges genre integrity but, at the same time, shows the real complexity of human behaviour. Hate propaganda can be found in a wide array of genres and media – e.g. postcards, posters, pamphlets, protest speeches, demonstrations, journal articles, parliamentary speeches, radio broadcasts, TV interviews, websites and social networks. The genres that are plausible candidates for hate propaganda are those originally constructed to ensure dissemination or influence public opinion (see section 4.1 of this chapter). Every genre whose territorial integrity is invaded by hate propaganda becomes a product of genre-bending, because its original generic features have been appropriated, exploited and manipulated by individuals or groups to accomplish a set of communicative purposes other than those for which the genre was originally constructed. Genre-bending is illustrated in the racist postcard in Figure 5.1, the anti-Semitic poster in Figure 5.2, the racist poster in Figure 5.3, and the racist literature in Figure 5.4. In each of these examples, the territorial integrity of the original genres – the postcard, the poster and the book – has been invaded and colonised by racist and anti-Semitic hate speech propaganda. The original genres function as instruments of deception because they lure the recipient to the hate group or organisation. The postcard (Figure 5.1) conveys hate against Chinese Americans through caricature and satirical text. The first poster (Figure 5.2) disseminates hate against the Jewish people, presenting them, through visual metaphor and caricature, as vermin threatening the world. The second poster (Figure 5.3) disseminates white supremacy by displaying a racist text and endorsing racial differences through a partisan comparison between the images of two young Germans and three Africans in stereotypical roles: the African savage, the servant, and the African American soldier. Under the appearance of scientific literature, the collection of books on display in Figure 5.4 disseminate and promote white supremacy.

On taking a closer look at the genres employed in hate propaganda, one can also observe the propensity for innovation prompted by technological change. Unlike in the past, when the speaker was more likely to gather a crowd in a public place and distribute posters and pamphlets, most hate propaganda today occurs online (Marwick & Miller 2014; Assimakopoulos, Baider & Sharon 2018; Lumsden & Harmer 2019; Winter 2019; Udupa, Gagliardone & Hervik 2021), typically on a blog, social media, website or other applications. Giltrow (2017) pointed to the capacity of new technologies to transform old genres into new ones by transposing them to new media or even providing bridges for genre formation when the new sphere of activity is consolidated (Herring, Scheidt, Bonus & Wright 2005). In addition, the digital age has given rise to new ways of disseminating propaganda worldwide. Algorithms are currently being used to create hate propaganda and spread it on social media, but they are also designed to detect it.[109]

5.2 System of genre, textual chain, intertextuality and interdiscursivity

Apart from genre-bending, other interesting concepts that can improve the understanding of the inherent complexity of hate propaganda are those of *system of genre*, *textual chain*, *intertextuality* and *interdiscursivity*. Bazerman (1994) introduced the concept of system of genre to refer to a limited range of "interrelated genres that interact with each other in specific settings" (Bazerman 1994: 97). This generic interaction is done to establish the current act concerning prior acts. I concur with Bazerman that:

> By understanding the genres available to us at any time, we can understand the roles and relationships open to us. Understanding generic decorum will let us know whether it is ours to ask or answer, argue or clarify, or declare or request (Bazerman 1994: 99).

For example, in a business transaction, the genre "placing an order" is established concerning prior acts expressed in generic form: giving a quotation, asking for a quotation, asking for samples on approval, making contact. Similarly, acknowledging receipt of the order, advising of despatch and making payment, amongst other limited range of genres, may appropriately follow upon placing an order to complete the business transaction. The concept of a system of genre throws light to the superstructure of the whole communicative event and the interaction between the parties involved.

---

109 In 2021, Naomi Nix and Lauren Etter drew attention to Facebook's inaction to stop hate speech on the social network (*The Print*, 25 October 2021). https://theprint.in/tech/facebook-knew-hate-speech-problem-was-bigger-than-it-disclosed-publicly/756379/ (accessed 31January 2022). In the same year, Elizabeth Dwoskin, Nitasha Tiku and Craig Timberg brought to light the biased algorithm used by Facebook to detect racist-motivated hate speech on its social network at the expense of Black users (*The Washington Post*, 21 November 2021). In their view, although researchers had proposed a fix to the biased algorithm, this was rejected by conservative partners. https://www.washingtonpost.com/technology/2021/11/21/facebook-algorithm-biased-race/ (accessed 31 January 2022).

The nature of hate propaganda does not establish a limited set of genres and acts that may, to a greater or lesser extent, appropriately follow in each situation. It is arguable then that the sequencing and consequences of acts expressed in generic forms may be harder to discern. The reason behind such loose structure lies in the fact that propaganda is not ascribed to any specific domain but, on the contrary, invades other domains and genres. The absence of a compact system of genre has important consequences for pragmatic analysis. As Bazerman (1994: 99) argues, since the illocutionary force-perlocutionary link is not preestablished, as it occurs in other specific-domain genres, it allows a wider array of perlocutionary effects (see Chapter 6 for further elaboration on the difficulty in prescribing the perlocutionary effects a hateful message may have).

Other useful concepts when analysing hate propaganda are those of textual chain, intertextuality and interdiscursivity. According to Reisigl and Wodak, a textual chain refers to:

> the sequence or succession of thematically or/and functionally related texts, which is pre-shaped by the frame of particular configurations of conventionalised linguistic practices (Reisigl & Wodak 2001: 79).

Text types in hate propaganda seem to establish intertextual and interdiscursive relationships between them that support the set of communicative purposes pursued by hate groups. Intertextual relations may be diverse: texts providing a context, texts within and around the text, texts referred to implicitly in the text, texts explicitly referred to in the text, texts embedded within the text, texts mixed with the text and quotations. Some hate groups follow a strategic transmission pattern to recruit people, indoctrinate them and incite hatred, hostility or violence against the members of a target group. This process may begin with a simple transmission, such as a profile or an online comment. Generally, these messages will contain instructions for obtaining more information via a website or any other application. The individual will then find pictures, texts, recordings and films, amongst other possibilities. The strategy is intended to initiate the individual from information recipient into the role of information seeker through reinforcement, and then from information seeker to opinion leader through indoctrination. It is through the transmission of hate speech that the seeds of hatred are planted in a person's mind. It is just a matter of time for the seeds to grow, and the person starts showing hostility and even, sometimes, performs violent acts against the targets.

Some of the court cases associated with hate speech in section 4.1 provide illustrative examples of textual chains, intertextuality and interdiscursivity. In Jersild v. Denmark (1994), the textual chain is made up of two journalistic genres that interact with each other: an article published in the serious press describing the racist attitudes of the Greenjackets in Østerbro (Copenhagen) and a television interview of some members of the same racist group, which contributed to the mass dissemination of their racist views. In A. v. The United Kingdom (2003), the textual chain consists of a parliamentary speech (political genre) on municipal housing policy delivered in the House of Commons and a press release (journalistic genre) of the same parliamentary speech. As in the preceding case, both genres interact, the press release contributing to the mass dissemination of the hate remarks embedded in the parliamentary speech. In United States v. Wilcox (2008), the textual change comprises profiles (digital genre) posted on an internet platform and a website (digital genre) associated with white supremacy ideologue David Lane. As in the other cases, both genres interact and foster the broad dissemination of racist messages. In ES v. Austria (2019), the textual chain includes a series of seminars entitled *Basic Information on Islam* (academic genre) and the website of a right-wing political organisation (digital promotional genre) that ensures the dissemination of such seminars.

On the other hand, in other cases, intertextuality and interdiscursivity are more implicit. For example, in both Brandenburg v. Ohio (1969) and Virginia v. Black (2003), the textual chain comprises a speech delivered by Klan leaders and the public display of burning crosses. The image of the burning cross evokes the discourse of racist violence against African Americans and two interrelated genres, Dixon's novel *The Clansman* (1905) and Griffith's film *The Birth of a nation* released in 1915, based on Dixon's referred novel (see Chapter 4).

In Fáber v. Hungary (2012), the textual change incorporates a racist demonstration of members of Jobbik and the display of the Árpád-striped flag at a site where Jews had been exterminated in large numbers during the Arrow Cross Regime. The image of the Árpád-striped flag at such a place may invoke anti-Semitism and the fascist discourse of the Hungarian Arrow Cross Party, a Nazi puppet government in place for seven months (October 1944–April 1945) in Hungary.

5.3 Hate propaganda as super genre

The concept of genre colony – that is, the grouping of closely related genres – serves two important functions in genre theory. First, it allows genres to be viewed at various levels of generalisation, making it possible to differentiate between super or macro-genres, genres and sub-genres. Second, it also makes it possible to relate these subcategories to contextual features (Bhatia 2014 [2004]).

Hate propaganda may be said to appropriate, manipulate and exploit "an entire repertoire of speech genres that differentiate and grow as the particular sphere develops and becomes more complex" (Bakhtin 1986: 60). In an attempt to bring some order to the apparent chaos of hate propaganda, I argue that the genres contained in the repertoire may be considered *sub-genres*. These can be grouped under the umbrella of a *super genre*: hate propaganda. The sub-genres form a genre colony because, despite their generic differences, they are geared to the same sphere of activity and share the same set of communicative purposes: to disseminate, advocate or incite hatred, hostility or violence against the members of a target group. The sub-genres are likely to differ in several other respects, such as their disciplinary affiliations, contexts of situation, and especially in their audience constraints (Tenor), topics (Field) and rhetorical conventions (Mode). Table 5.1 below depicts the genre colony represented by hate propaganda. Far from being exhaustive, the genre list below is meant to give a general idea of hate propaganda as a dynamic, innovative and versatile super genre that crosses disciplines, genres, means of expression and media.

Table 5.1: Hate propaganda as super genre: Colony of hate speech genres.

- **Written genres**
  Postcard
  Poster
  News item
  Academic or professional publication
  Legal norm, statute, act
  Lyrics
  [. . .]
- **Visual genres**
  Caricature
  Photograph
  Picture
  Graffiti
  [. . .]
- **Oral genres** (onsite or broadcast through mass media)
  Protest speech at a rally
  Parliamentary speech
  Interview
  [. . .]

Table 5.1 (continued)

– **Audiovisual genres**
Film
Performance
Video
[. . .]
– **Internet genres**
Text messages
Blog
Tweet
Website
Social networks conversation
[. . .]

5.4 The genre integrity of hate propaganda

According to Bhatia, *genre integrity* refers to

> a socially constructed typical constellation of form-function correlations representing a specific professional, academic or institutional communicative construct realising a specific communicative purpose of the genre in question (Bhatia 2014 [2004]: 142).

A genre, then, has been successfully constructed if it serves the communicative goals for which it was constructed. I claim that, in the case of hate propaganda, the super genre has generic integrity, but that this has a special nature for varied reasons. First, it is complex because it reflects genre-bending, resulting in hybridisation, mixing, or embedding of two or more generic forms. Second, it is dynamic because it reflects a gradual development over the years prompted by the advent of new technologies. Third, it is versatile because it can accommodate its communicative purposes to the given circumstances. Fourth, it has a recognisable generic character for the members of the speech community.

5.5 The power of hate propaganda

According to Bhatia (2014 [2004]), academic and professional communities can express their communicative intentions and meet their goals by using the same set of genres repeatedly and over time. The recurrent use of a range of genres contributes to "solidarity within its membership, giving them their most powerful weapon to keep outsiders at a safe distance" (Bhatia 2014 [2004]: 223). Although

Bhatia discussed the *power of genre* in academic and professional discourses, I attempt to demonstrate that that power can be equally relevant in the context of hate propaganda.

As a super genre, hate propaganda contributes to building cohesion within the members of the ingroup. As in the case of academic and professional communities, the power of genre in a hate group involves knowledge that is accessible only to its membership. This knowledge comprises both text-internal and text-external aspects (cf. Bhatia 2014 [2004]: 147–151).

On the one hand, text-internal aspects relate to knowledge about the appropriate register (the context of situation, lexico-grammar and rhetorical conventions) and intertextuality. On the other hand, text-external aspects concern three types of knowledge: (1) *genre appropriacy*, (2) *genre construction* and (3) *disciplinary culture*. First, knowledge about genre appropriacy allows the hate group to choose the most convenient genres for hate propaganda. Second, knowledge about genre construction allows experienced members of the hate group to produce, use and interpret hate propaganda conveniently. Third, knowledge about disciplinary culture allows experienced members of the hate group to exploit and manipulate generic conventions to express the group's communicative intentions within the context of specific social activities. As Bhatia (2014 [2004]) claimed, the power of genre also relates to the capacity to innovate and create novel generic forms. One of the most noticeable characteristics of hate propaganda, as earlier mentioned, is its dynamism and capacity to innovate.

Genre knowledge must be shared between the producers of hate propaganda and the intended audience. In other words, for hate propaganda to be successful, the generic forms used to express the hateful messages must match the intended audience's expectations. This match is possible only when all the participants in the speech event share the code and the genre knowledge, including the knowledge about its construction, use and interpretation. Whereas the shared genre knowledge builds cohesion and homogeneity within the ingroup, it creates a social distance between those considered legitimate members of the hate group (the ingroups) and those considered outsiders (the outgroups or targets). This social distance may result in negative consequences for the outgroups because they do not have access to the necessary shared background knowledge and, thereby, may be unable to defend themselves from verbal assault or bring a claim against the perpetrators. In addition, the triers of fact may fail to interpret the hateful messages appropriately, as they may not always be aware of the linguistic resources and tools used to create hate propaganda.

6 Conclusions

This chapter analysed the texts in which hate speech is likely to manifest itself, their register and complex generic forms. It was explained that hatred must be linguistically encapsulated in texts that make hateful expressions and acts durable over time. As a result, the hate text production situation and the hate text reception situation may not be adjacent in space and time. Thus, the imminence standard (see Chapters 2 and 3) may be, in effect, difficult to assess, especially after the advent of online hate speech.

After analysing hate speech through the lens of register theory, I conclude that, in theory, a hate register may be described through the elements of the context of situation: Tenor, Field and Mode. Specifically, the Tenor may be indicative of a hate register when the participants of the speech event are: (1) the hate-advocating speaker (the saviour), (2) the ingroup (the victims), (3) the outgroup (the dangerous threat) and (4) the overhearers, bystanders or eavesdroppers. The Field may indicate a hate register when the text contains, amongst other things, fallacies about the moral inferiority or non-humanity of the outgroups and false statements of fact harming their dignity and social reputation. Although the rhetorical mode of a hate text may be characteristically manipulative, it is important to note that its forms of expression are less definite because of the varied channels of communication and forms the hate-advocating speaker may appropriate, manipulate and exploit. A hate register may also be described for the set of malicious communicative purposes it pursues. This set of communicative purposes may not always be easily determined when conveyed implicitly and indirectly (see Chapters 6 and 8). As a result of the increasing criminalisation of hate speech worldwide, the hate-advocating speaker will tend to convey hate speech implicitly and indirectly. Indirectness evades automatic computer detection even though it is recognisable by the audience it addresses. In looking at the success rate of natural processing (NLP) in detecting hate speech, Knoblock (2022: 4) claims that NPL still has a lower success rate (60%) when the messages are covert than when they are overt (80%).

Hate groups can appropriate, manipulate and exploit wide-ranging ideologically-based linguistic features, and use them as verbal weapons to propagate hate and instigate hostility or violence towards the members of a target group. In this chapter, several arguments were tabled to reject the hypothesis that a hate register may be described for the pervasive (statistically significant) linguistic features in the text. First, proposing a closed catalogue or inventory of pervasive linguistic features may be fruitless because these may vary from one genre to another. Second, it may be scientifically inaccurate to establish a correlation between a word's high frequency of occurrence in a text with its status as a hate register linguistic feature because words not present in a high-frequency list may be as relevant as those present (cf. Brindle 2016; Ruziate 2018; Technau 2018).

By contrast, the present author supported the idea that the collective salience of a linguistic feature seems to be a more accurate indicator of a hate register. A linguistic feature may have collective salience when it is prominent – that is, when it has the power to invoke social prejudice, hostility or violence against a target group. Such power derives from the ideas, beliefs, views and attitudes the speaker shares with the intended audience (the ingroup). Arguably, the collective salience of the linguistic features of a hate register may be sometimes invisible in the eyes of the law unless the judge or jury is familiar with the particular hate group register. Consequently, hate-advocating speakers may elude legal responsibility and profit from impunity.

This chapter also analysed hate speech through the lens of genre theory. Hate speech cannot be ascribed to a pure genre identifiable by specific rhetorical conventions or textualised patterns. The problem of generic adscription makes hate speech recognition and interpretation difficult, especially for those who do not belong to the same speech community. By contrast, from a wider social perspective, hate speech may, in effect, be associated with a specific sphere of negative activity that has become characteristic of it, giving rise to what we know as hate propaganda. It was argued that hate propaganda might be categorised as a super genre encompassing an entire repertoire of hybrid sub-genres (a genre colony). Hate propaganda illustrates the phenomenon of genre-bending, which would not be possible if it were not for the fact that hate-advocating speakers have both text-internal and text-external knowledge. This knowledge helps them select the most convenient domains and genres to achieve their malicious communicative purposes.

Another interesting insight given by genre theory is that the superstructure (the system of genre) of hate propaganda is not preestablished. Therefore, the illocutionary force-perlocutionary link between the acts expressed in generic forms and their perlocutionary effects is unpredictable (see Chapter 6). Despite the loose system of genre exhibited by hate propaganda, I argued that it has genre integrity, although complex due to its dynamism and versatility.

6 Speech act theory

1 Introduction

Language users identify examples of hate speech by the types of words people use, the contents people express with words and the acts people do with words. Doing things with words is at the heart of Speech act theory (Austin 1962; Searle, Kiefer & Bierwisch 1980). In essence, this theory explains that words can be used to present propositional content and perform actions in the real world. Words can, in effect, be employed to promote peace, solidarity and understanding or, by contrast, be used as verbal weapons, as is the case of disinformation, propaganda and hate speech. This chapter looks at hate speech through the lens of Speech act theory. One major concern for both linguists and legal practitioners is how to discern which speech acts are classifiable as hate speech. This task is not easy, as the reader will see in this chapter.

Speech act theory has been criticised for its philosophical foundation, which is concerned with the conceptual truths that underline any possible language or system of communication (Searle 1979: 162). Some linguists, such as Levinson (1983), Holtgraves (2005) and Sadock (2007), have pointed to the difficulties Speech act theory encounters when analysing empirical facts of natural human languages. Given such criticisms and concerns, in this chapter, my aim is to determine to what extent Speech act theory may explicate the speech acts giving expression to hate speech. To begin with, in the following sections, I will analyse the speech acts that may be considered surface manifestations of hate speech by using the tools provided by the theory.

2 Hate speech: A sequence of speech acts

The laws regulating the prohibition of hate speech are designed to keep this social scourge from escalating into something more dangerous, such as unlawful acts. International law (see Chapter 2) prohibits three main acts: *dissemination*, *advocacy* and *incitement* to hatred, hostility or violence against legally-protected groups. Apart from these three main acts, there are other acts prohibited by law, such as *encouragement*, *promotion*, *justification* and *harassment*. Nevertheless, these acts are not so frequently included in technical legal definitions. According to van Dijk (1992), describing the pragmatic structure of discourse means giving an account of its cohesion: how the speech acts are organised linearly in sequences and hierarchically in global speech acts. A discourse is coherent when all the speech acts

https://doi.org/10.1515/9783110672619-006

accomplish one global speech act. It could then be arguable that dissemination, advocacy and incitement are not unrelated speech acts but, instead, are components of a *sequence of speech acts* (van Dijk 1992) organised linearly and hierarchically because they can together be assigned one global intention (Bukhardt 1990) or plan: to engage the audience in unlawful acts against the targets. Besides, the component speech acts in the sequence can also be classified according to their contribution to the realisation of another or other acts in the sequence (cf. van Dijk 1992):

a) Dissemination can be described as a *preparatory speech act* because its result is a necessary condition for the success of advocacy and incitement.
b) Advocacy can be described as an *auxiliary speech act* because the success of its result is a sufficient condition for the success of incitement.[110]
c) Incitement can be described as the main act of the sequence.
d) The audience's engagement in unlawful or violent acts may be described as a consequent act (perlocutionary act) derived from the main act's performance.

It is important to note that the component speech acts in the sequence are major acts because other minor acts can support them. Figure 6.1 shows how the described structure of speech acts accounts for the linear and hierarchical organisation of hate speech as a global speech act.

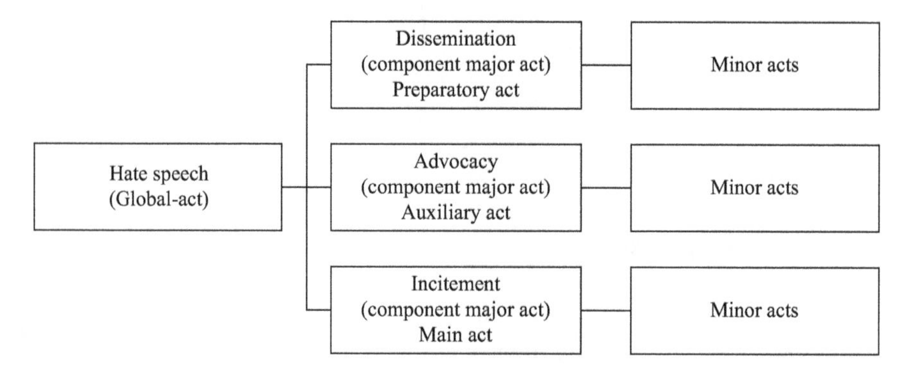

Figure 6.1: Hate speech as a sequence of speech acts at a macro level.

110 At a macro level, other speech acts such as promotion, justification and encouragement may be classified as auxiliary acts to the main act in the sequence. Harassment, however, may be considered a separate category because it is a consequent act (a perlocutionary act) that may result from the felicitous performance of the main act: incitement to hatred, hostility or violence towards or against the targets.

3 Hate speech: A taxonomy of illocutionary acts

One important concern within Speech act theory is the classification of speech acts. This section considers Searle's speech act classification (1969; 1979: 1–29), which is by far the one most widely used by linguists, and explores the insights it may provide into hate speech. Searle's classification was based on pragmatic parameters known as *direction of fit*, i.e. whether the words fit the world's facts or the world comes to fit the words. Specifically, Searle's taxonomy consists of five well-known categories:

1) *Representatives* (words-to-world direction of fit): These are acts that commit the speaker to the truth of the expressed proposition – e.g. assert, conclude, complain, deduce, amongst others.
2) *Directives* (world-to-words direction of fit): These are acts that reflect an attempt by the speaker to get the hearer to do something (in varying degrees) – e.g. ask, order, command, request, instruct, urge, warn, advise, encourage, incite, amongst others.
3) *Commissives* (world-to-words direction of fit): These are acts that commit the speaker (in varying degrees) to some future course of action – e.g. promise, threaten, offer, undertake, swear, guarantee, amongst others.
4) *Expressives* (no direction of fit): These are acts that express a psychological state specified in the sincerity condition about a state of affairs stated in the propositional content – e.g. thank, congratulate, regret, deplore, apologise, detest, amongst others.
5) *Declarations* (words-to-world direction of fit and world-to-words direction of fit): These are acts that bring about immediate changes in the institutional state of affairs; hence, they tend to rely on extra-linguistic institutions – e.g. declare, appoint, nominate, sentence, pronounce, amongst others.

Furthermore, Searle (1969: 66–67) proposed four felicity conditions that must be met for the above-mentioned types of acts to be performed successfully:
Rule 1: Propositional content. This rule establishes the semantic meaning of a sentence that has been uttered.
Rule 2: Preparatory conditions. These conditions are concerned with the speaker's and hearer's attitudes, expectations and preferences towards the truth of the propositional content. They also encompass the lexical meaning of performative verbs.
Rule 3: Sincerity conditions. They are related to the speaker's and hearer's plans and communicative intentions.
Rule 4: Essential conditions. They summarise the result of the successful speech act.

Assigning an act to a single category may not always be easy. For instance, as Searle (1975) noted, there may be a certain overlap between representatives and declaratives and also between directives and commissives. In the following, I will determine the types of speech acts expressing hate speech under Searle's taxonomy.

3.1 Hate speech: An expressive?

Stating that hate speech is an *expressive* act because it expresses hatred may not be accurate, as Brown (2017a) explained in his essay on the myth of hate. Hate speech might, in effect, be driven by a range of emotions and motives other than hatred. For example, hate speech may result from anxiety, anger, frustration, or fear of loss of social privileges caused by the presence of any foreigners in the social community; it may also obey other motives such as attention-seeking, pleasure in being controversial or economic self-interest. In addition, if hate speech were reduced to the expression of hate (*I hate* [. . .]) or other emotional content, the law would invade the terrain of freedom of speech because it would prohibit important forms of self-expression and self-realisation. Human beings also need to release negative feelings and thoughts through speech. Unlike in totalitarian states, in democratic societies, laws cannot be used to control people's thoughts, beliefs and attitudes, no matter how pernicious these may be.

I agree with Brown (2017a) in that hate speech is motivated by hatred, but this motivation is irrespective of whether or not the utterance serves to express that hatred. Hatred is the fuel needed to provoke violent action. When racist speakers perform hateful actions, they arouse emotions or attitudes of hatred in the intended audience that can lead to violence. Let me give some substance to this by taking a couple of examples. The first one (Example 6.1) is a hate remark posted on 10 May 2021 online in the web-forum *Proud Boys Uncensored*.[111]

111 The Proud Boys are a far-right, neo-fascist, exclusively male organisation that promotes and engages in political violence in the United States. Some members of The Proud Boys have been indicted for conspiracy related to the 6 January 2021 attack on the United States Capitol. When this book was written, The Proud Boys had not yet been banned from Telegram's social network. https://web.telegram.org/#/im?p=@TheWesternChauvinist. The message was retrieved from the account "Proud Boys Uncensored". https://t.me/proudboysuncensored (accessed 10 May 2021).

Example 6.1
Black parasite brutally punches white woman. This is what happens when our women are propagandised to hate our men. They end up getting abused and beaten by alien invaders.

The above message includes a racial epithet ("Black parasite"), a slur ("alien invaders) and a fallacy – that is, faulty reasoning in the construction of the argument related to the incident. Although the message is motivated by hatred, it may not express hatred. Other motives other than hatred, such as anger or fear of loss of socio-economic power caused by the presence of foreigners in the social community, may explain the speaker's behaviour towards immigrants. It may be said that this post performs two speech acts: the speaker detests immigrants (expressive) and indirectly incites hatred towards them (directive).

The second example (Example 6.2) is taken from the website of the Westboro Baptist Church (WBC).[112]

Example 6.2
Sodomites are wicked and sinners before the Lord exceedingly (Gen. 13:13), they are violent, and they doom nations (Gen. 19:1–25; Jgs. 19), they are nasty and abominable (i.e., exceedingly filthy) in the sight of God, whom Himself says that they are worthy of death for their vile sex practices (Lev. 20:13; Rom. 1: 32).

Example 6.2 reproduces a hate remark including several slurs, such as "sodomites", "wicked", "sinners", "nasty", "abominable", "exceedingly filthy" and "worthy of death". The remark maliciously re-contextualises excerpts from the Bible to vilify homosexuals. The message is hate-fuelled, but the speaker may be driven by a motive other than hatred, for instance, religious fanatism. As in the case of Example 6.1, this post may perform two speech acts: the speaker detests homosexuals (expressive) and indirectly incites hatred towards them (directive).

The existence of speech acts motivated by hatred or speech acts capable of stirring up emotions, feelings or attitudes of hatred invites reflection about Searle's expressive speech act category, which fails to account for the complexity of human linguistic behaviour.

112 The Westboro Baptist Church (WBC) is an American hyper-Calvinist hate group. It is known for engaging in hate speech against Jews, Muslims, homosexuals and transgender people. https://www.godhatesfags.com/ (accessed 20 March 2022).

3.2 Hate speech: A directive?

Hate speech is regulatable not because it expresses hate or is motivated by hatred, but because it can incite unlawful or violent acts against members of groups identifiable by legally-protected characteristics. Hate speech may, then, fall within Searle's category of directives because it conveys an attempt (in varying degrees) by the speaker to get the addressee to perform some future unlawful or violent action against the targets. The directive can be formulated as *I want you to engage in unlawful or violent action against the targets*. A directive must accomplish certain felicity conditions to guarantee its successful performance. The felicity conditions that incitement must meet are similar, to a more or lesser extent, to those of other directives such as order or demand (Searle 1969: 36):

Propositional content: Future A by H.[113]
Preparatory conditions: (i) S believes A needs to be done, (ii) H can do A, (iii) H has no obligation to do A[114] (iv) S has the right to tell H to do A.
Sincerity condition: S wants H to do A.
Essential condition: Counts as an attempt to get H to do A.

It might be noted that in a directive like demand or order, the Preparatory conditions, particularly (iii), implies that H must do A. By contrast, in the case of incitement, H has no obligation to do A. This discrepancy would explain why hate-advocating speakers must do strategic work to persuade their audience. Hence, incitement is more likely to be performed implicitly and indirectly by other acts, instead of directly and explicitly, as the following section shows.

4 Hate speech: Explicit and implicit performatives

Austin (1962) used the term *performative* to describe utterances that explicitly convey the action named by the verb – e.g. promise, threaten, demand and refuse, while he used the term *constative* to refer to a class of fact-stating utterances, which establish that something is true or false – e.g. state, describe, assert and predict. It is well-known that Austin changed his view in later work because he attributed performativity to all utterances and verbs. This relevant insight led

113 Following Searle (1969), S stands for Speaker, H stands for Hearer and A stands for action.

114 Note that one of the preparatory conditions of a directive speech act such as demand implies that H (Hearer) must do A (Action).

Austin to make a distinction between *explicit* and *implicit performatives*, based on the assumption that most speech acts do not exhibit an overt performative verb. Explicit performatives (Recanati 1980: 205–220) are relatively clear because they include the performative verb. For example, in the utterance *I promise I will call you when I arrive*, the verb *promise* is an explicit performative because it names the speaker's communicative intention – the illocutionary force of the utterance (Austin 1962). Similarly, Thomas (1995) categorises explicit performatives as metalinguistic acts because the verb by itself points to what type of action the speaker is doing in words – e.g. in *I advise you*, the speaker is explicitly advising the hearer. Thus, speech act recognition is relatively unproblematic in the case of explicit performatives, as they unambiguously contain the speech act the utterance conveys. The use of performative verbs may be constrained by specific circumstances – e.g. a situation in which the speaker wants to convey a formal tone, clarify intention or avoid ambiguity (Holtgraves 2005: 2041).

In real conversation, utterances often include implicit performatives – that is, speech acts that do not exhibit the performative indicating the utterance's illocutionary force. Let us take, for instance, the explicit performative *promise*. If a speaker wants to promise something, they can do it either with an utterance using an explicit performative: *I promise I will call you when I arrive*, or an implicit performative, as in *I will call you when I arrive*. Given the same context of situation, the hearer is likely to understand, without any additional difficulty, the same speaker's communicative intention (illocutionary force) in both cases, although the latter utterance leaves the speaker's communicative intention open to interpretation.

Austin (1962) proposed that explicit and implicit performatives are subject to felicity conditions. These rules specify when an utterance is felicitous or appropriate for a particular occasion. Austin (1962: 14–15) distinguished three main categories of felicity conditions:

A.1 There must exist an accepted conventional procedure having a certain conventional effect, that procedure must include the uttering of certain words by certain persons in certain circumstances, and further,

A.2 the particular persons and circumstances in a given case must be appropriate for invoking the particular procedure.

B.1 The procedure must be executed by all participants both correctly and

B.2 completely.

C.1 Where, as often, the procedure is designed for use by persons having certain thoughts or feelings or for the inauguration of certain consequential conduct on the part of any participant. A person participating in and so invoking the procedure must have those thoughts or feelings, and the participants must intend to so conduct themselves and further

C.2 must so conduct themselves subsequently.

When one or more of these conditions are not met, Austin (1962) explained that this failure may result in one or more of the following three infelicities:

1) *Misinvocations*: These occur when any of the A conditions above are not met and hence, the purported act is disallowed.
2) *Misexecutions*: These occur when any of the B conditions above are not met, and hence the purported act is vitiated by errors or omissions.
3) *Abuses*: These refer to those situations in which the act succeeds but is hollow and, therefore, does not meet the C conditions because the participants do not have the expected commitment or feelings associated with the felicitous performance of the act in question.

Engaging in hate speech is not merely to speak; it is also to perform a speech act, participate in a social practice of discrimination, and reproduce cognitive elements associated with social prejudice and intolerance. To emphasise the performative nature of hate speech, some legal scholars use the term *assaultive speech* (Brown 2017b: 31) as an alternative to hate speech.

Despite the absence of a unified technical legal definition of hate speech, this is commonly associated with a sequence of macro speech acts: disseminate, advocate and incite hatred, hostility or violence against members of groups identifiable by legally-protected characteristics. Paradoxically, disseminate, advocate and incite are explicit performatives that are never performed explicitly in hate speech: *I disseminate* [. . .], *I advocate* [. . .], *I incite* [. . .]. The hate-advocating speaker's failure to use explicit performatives is not associated with an ill-formed communicative intention or face-saving purposes (see Chapter 7). Far from that, the explicit performatives at the core of hate speech are always expressed through implicit performatives – and quite often through indirect speech acts too, as shown in the next section. The reason for this strategic choice is simple: explicitness may result in the closure of the platform from where the messages are sent, indictment or even criminal prosecution. By way of illustration, let us consider two examples associated with hate speech. Example 6.3 is an excerpt from Terminiello v. Chicago (1949), to which the author has already referred in the preceding chapters.

Example 6.3

It is the same kind of tolerance, if we said there was a bedbug in bed, "We don't care for you" or if we looked under the bed and found a snake and said, "I am going to be tolerant and leave the snake there." We will not be tolerant of that mob out there. We are not going to be tolerant any longer (Terminiello v. Chicago, 337 US 1 (1949), p. 337 U.S. 21).

In Example 6.3, the participants of the speech event were the speaker (Terminiello) and his audience, which can be divided into two subgroups: (1) the audience inside the auditorium (the ingroups) and (2) the protesters outside the auditorium (the outgroups). In Terminiello's speech, there is no single utterance including the explicit performative: *I incite you to act violently against Jews and communists.* However, Terminiello was judged by the Chicago Court for having abused his right to freedom of expression by uttering fighting words that infringed the Chicago's breach of the peace ordinance. Fighting words include:

> the lewd and obscene, the profane, the libellous, and the insulting or *fighting words* – those which by their very utterance inflict injury or tend to incite an immediate breach of the peace (Chaplinsky v. New Hampshire, 1942).[115]

The above quote implies that fighting words may convey incitement to violent acts implicitly. Clearly, in Terminiello's inflammatory speech, the explicit performative *I incite* [. . .] was replaced with fighting words, such as the slurs "bedbug" and "snakes". Terminiello's speech was motivated by hatred, and this hatred triggered incitement to violence (directive speech act). In comparing Jews and communists to parasites and reptiles, Terminiello dehumanised the targets (Stollznow 2017). Dehumanisation is a psychological weapon that strategically enables members of dominant groups to morally disengage from the suffering of the disadvantaged group, thereby facilitating acts of intergroup violence (cf. Bruneau & Kteily 2017).

Although the United States Supreme Court overturned Terminiello's guilty verdict, his implicit act of incitement to violence was felicitous from the perspective of Speech act theory. The act accomplished Austin's felicity conditions: Terminiello belonged to a dominant group and had the religious authority to deliver his speech (A category); Terminiello was an experienced language user who knew how to provoke hatred in the audience and led them to engage in violent intergroup action (B category). Lastly, all the participants in the speech event (the ingroups and the outgroups) had the expected commitment or feelings associated with the act (C category).

Example 6.4 is an excerpt from Brandenburg v. Ohio (1969), to which the author has also referred in the preceding chapters.

115 Fighting-words jurisprudence began in the 1942 decision Chaplinsky v. New Hampshire. https://caselaw.findlaw.com/us-supreme-court/315/568.html (accessed 20 May 2022).

Example 6.4

This is an organiser's meeting. We have had quite a few members here today which are – we have hundreds, hundreds of members throughout the State of Ohio. I can quote from a newspaper clipping from the Columbus, Ohio, Dispatch, five weeks ago Sunday morning. The Klan has more members in the State of Ohio than does any other organisation. We're not a revenge organisation, but if our President, our Congress, our Supreme Court, continues to suppress the white, Caucasian race, it's possible that there might have to be some revengeance taken (Brandenburg v. Ohio, 395 US 4444 (1969), p. US 446).

In Example 6.4, the participants in the speech event – a Klan brotherhood rally – were the speaker – Brandenburg, a Klan leader – and the audience – Klan members. Because the rally was broadcast, Brandenburg's speech also had wide dissemination. The selected excerpt illustrates an implicit threat. It might be noted that Brandenburg did not perform a commissive explicitly: *I threaten X to do Y if X does Z*", but, instead, implicitly through some thought-provoking moves.

Although the implicit threat was conventionally conveyed with a conditional sentence: *If [. . .] then [. . .]* (Holtgraves 2005; Muschalik 2018), Brandenburg did not formulate a clear first conditional expressing absolute probability: "If our President, our Congress, our Supreme Court, continues to [Present tense] suppress the white, Caucasian race, we will [Future tense] take revenge". By contrast, Brandenburg expressed the act of threatening implicitly and ambiguously, probably with the intention of avoiding indictment and criminal prosecution. Significantly, the Klan leader chose a mixed conditional. Whereas the condition was uttered in the present tense, expressing absolute probability, the consequence was strategically designed to hide the speech act's illocutionary force, at least for those who do not belong to the ingroup. Firstly, Brandenburg introduced the main clause through a complex syntactic pattern: *it's + modal adjective + that clause (focus).* His purpose was to express certainty about the following clause using focus: "it's possible that there might have to be some revengeance taken." Note that Brandenburg cleverly tried, at the same time he uttered probability, to elude responsibility by using "but" as a disclaimer (cf. Geyer, Bick & Kleene 2022: 241–261). The disclaimer contributes to the overall strategy of positive self-presentation because the clause introduced by "but" is always negative about the Others. Apart from the disclaimer, Brandenburg used other impersonal structures (cf. Knoblock 2022: 2), clearly eluding self-responsibility in the act, such as *there* and the *passive voice* in "have to be some revengeance taken". In addition, Brandenburg hedged the degree of certainty by including the modal "might" in the consequence clause.

Although Brandenburg's conviction was reversed by the Supreme Court, from the perspective of Speech act theory the implicit threat was felicitous. According to Austin's (1962) felicity conditions: Brandenburg belonged to a dominant group and was an authorised speaker to deliver the speech (A category). Brandenburg was an experienced language user and knew how to use language strategically to cause violent actions (B category). The participants in the speech event (the Klan members) had the expected commitment or feelings associated with the act (C category).

In sum, it has been shown that the speech acts that are surface manifestations of hate speech are likely to be performed implicitly and indirectly, and can be linguistically encapsulated in varied forms.

5 Hate speech: Direct and indirect speech acts

Austin's (1962) implicit performatives left the door open to Searle's (1969; 1979: 30–57) division between *direct* and *indirect speech acts*. An utterance is seen as a direct speech act when there is a direct relationship between the utterance's form and function, whereby the act explicitly indicates the speaker's intended meaning. Some typical examples are:

> A declarative is used to make a statement: *It is cold in here.*
> An interrogative is used to ask a question: *Don't you feel cold?*
> An imperative is used to make a command: *Please close the window!*

On the other hand, an utterance is seen as an indirect speech act when there is an indirect relationship between the utterance's form and function; hence, the act implicitly indicates the speaker's intended meaning. For example, a declarative sentence – e.g. *It is cold in here* – can be used to make a request – e.g. *Could you, please, switch on the central heating?* Any competent speaker can perfectly understand that, in a given context, the illocutionary force of a declarative sentence like *It is cold in here* is not to inform about a room's low temperature but to make an indirect request.

Searle (1969) introduced the notions of *primary* and *secondary illocutionary acts* in his analysis of indirect speech acts. In Searle's view, the primary act of an utterance like *It's cold in here* is indirect (a request), and the secondary act is direct (a statement about the room's temperature). The co-occurrence of primary and secondary acts in the same utterance allows speakers to understand how it is possible for someone to say something and mean it but at the same time mean something else. Searle (1975) laid out the logical steps by which the hearer of an indirect speech act figures out the speaker's intended meaning. As an illustration, consider Example 6.5.

Example 6.5
A: It's cold in here.
B closes the window.

The interpretative process can be broken down into at least five unconscious steps:
Step 1. A statement is made by A, and B responds non-verbally.
Step 2. The literal meaning of A's utterance is not relevant to the conversation.
Step 3. Since B assumes that A is cooperating, there must be another meaning to the utterance.
Step 4. Based on mutually shared background information, A knows that B is closer to the window.
Step 5. B knows that A has intended something other than the literal meaning and that the primary illocutionary act must have been a request.

Thus, what superficially looks like a statement about the room's temperature is primarily a request. The lack of mapping between an utterance's form and function implies that one needs to consider the speech acts independently of the utterances used to perform them.

Indirectness is, in effect, very frequent in actual verbal interactions, especially due to face-saving work (see politeness strategies in Chapter 7). It is very common to find indirectness in hate messages, as a speaker's strategic move to hide an utterance's illocutionary force with ambiguity. To illustrate the role indirectness plays in hate speech, let me go back to Terminiello v Chicago (1949) in Example 6.3. For the Chicago court, the relevant speech act in the case was incitement to violence. As the reader will see in the following, this directive was conveyed indirectly by other speech acts.

Firstly, Terminiello made an assertion, a speech act that falls under the category of representatives:

> It is the same kind of tolerance, if we said there was a bedbug in bed, "We don't care for you" or if we looked under the bed and found a snake and said, "I am going to be tolerant and leave the snake there."

Although Terminiello's assertion may be felicitous for the ingroup, it is infelicitous from the perspective of Speech act theory. According to Searle (1969), an assertion must meet the following conditions for its guarantee to be felicitous:

Propositional content: Any proposition p.
Preparatory conditions: (i) S has evidence for the truth of p, (ii) It is not obvious for S and H that H knows.
Sincerity condition: S believes p.
Essential condition: Counts as an undertaking to the effect that p represents an actual state of affairs.

It might be noted that Preparatory condition (i) – S has evidence for the truth of p – is not met in Terminiello's assertion. The analogy between Jews and communists and "bedbugs" (parasites) and "snakes" (evil) is based on false reasoning, because it is not supported by scientific evidence. Because of this, Terminiello's assertion may be said to incur a fallacy of inconsistency with the result that it is infelicitous from the perspective of Speech act theory. It is likely that the ingroups did not perceive such comparison as an inconsistency. For the ingroups, portraying Jews and communists as "bedbugs" (parasites) and "snakes" (evil) has collective salience (see Chapters 5 and 8) and reinforces their social prejudice and intolerance towards the targets.

Secondly, it might also be argued that comparing Jews and communists to "bedbugs" and "snakes" is an expressive act because it conveys the speaker's aversion towards the targets.

Thirdly, Terminiello's utterance also conveys incitement to hatred, hostility or violence through commissives. A commissive is described as an attempt by S to get H to believe that S is committed to a particular course of action:

We will not be tolerant of that mob out there.
We are not going to be tolerant any longer.

At a surface level, the above utterances describe a particular type of commissive: promise with certainty or guarantee. This act can be felicitous if it meets the following conditions (Searle 1969):
Propositional content: Future action A by S.
Preparatory conditions: (i) S believes H wants A done; (ii) S can do A; (iii) A has not already been done; (iv) H benefits from A.
Sincerity condition: S is willing to do A.
Essentially condition: Counts as an attempt by S to make H believe about A to be done by S.

As in the case of directives, the propositional content of a commissive specifies a future course of action, but in this case, it is an action of the speaker rather than the hearer. Since Terminiello used the first-person plural – "We" – he, an authorised speaker, was speaking on behalf of the ingroup. According to Thomas' taxonomy (1995), such an act could fall into the *group performatives* category, because the act is not felicitous unless performed by the group.

The case of Jersild v. Denmark (1994), discussed in Chapter 3, illustrates the simultaneous presence of dissemination, advocacy and incitement to hatred, hostility or violence in the same case. For the utterance of certain racist remarks in a broadcast interview, the Supreme Court in Denmark condemned some members of the Greenjackets in Østerbro (Copenhagen) for racist speech, while both the journalist who conducted the interview and the head of the news section were convicted for disseminating racist ideology. Later, the majority vote of the European Court of Human Rights overturned the conviction of the journalist and the head of the news section, giving more weight to freedom of the press than to the protection of minorities whose human dignity had been attacked. In this case, the Greenjackets incited hatred (directive) indirectly by making several racist remarks (representative), denying Black people their humanity and uttering racial epithets ("nigger") and other slurs that express aversion to immigrants (expressive).

Example 6.6

The Northern States wanted that the niggers should be free human beings, man, they are not human beings, they are animals.

Just take a picture of a gorilla, man, and then look at a nigger, it's the same body structure and everything, man, flat forehead and all kinds of things.

A nigger is not a human being, it's an animal, that goes for all the other foreign workers as well, Turks, Yugoslavs and whatever they are called (Jersild v. Denmark. Application No. 15890/89. Judgment. Strasbourg. 23 September 1994, p. 9).

As in the case of Terminiello's assertion, the above remarks were felicitous for those who share the same racist views – e.g. a racial epithet like "nigger" has the power of invoking racism and xenophobia (Hom 2008). By contrast, such remarks were infelicitous from the perspective of Speech act theory: The Preparatory condition (i) – S has evidence for the truth of p – is not met. The remarks are based on a fallacy of inconsistency – that is, the analogy between a Black person or an immigrant with an "animal" or a "gorilla" is false because it is not scientifically grounded. It might also be arguable that through the expression of racial epithets and slurs, the speakers are expressing intense dislike, loathing and hostility towards immigrants.

6 Hate speech: A complex act

From the above discussion, it is obvious that hate speech is not a single act but a *complex act* (Bruno, Bosco & Bucciarelli 1999) at a micro-level (Figure 6.2). I argue that hate speech can be described as a complex act because it involves two dimensions: (1) what speakers commit themselves to and (2) what they call on the hearer to perform (Beyssade & Marandin 2006). I will try to give some substance to my claim in the following. Hate speech is a complex act because the main act, incitement to hatred, hostility or violence, must be performed indirectly through other acts. Therefore, when the speaker pronounces an utterance, they can say one thing and mean it while at the same time they mean another thing. The hearer's task is to infer which is the primary illocutionary act and which is the secondary illocutionary act in a given context. In the case of hate speech, the complex act consists of a directive (incitement to hatred, hostility or violence) as the primary illocutionary act. The secondary illocutionary acts can be of different types – e.g. representatives, directives, commissives and expressives. It is noteworthy that expressives are usually embedded in other acts, their role being to provide the necessary fuel to incite hatred, hostility or violence (the primary illocutionary act). The context of situation provides the necessary cues to adequately

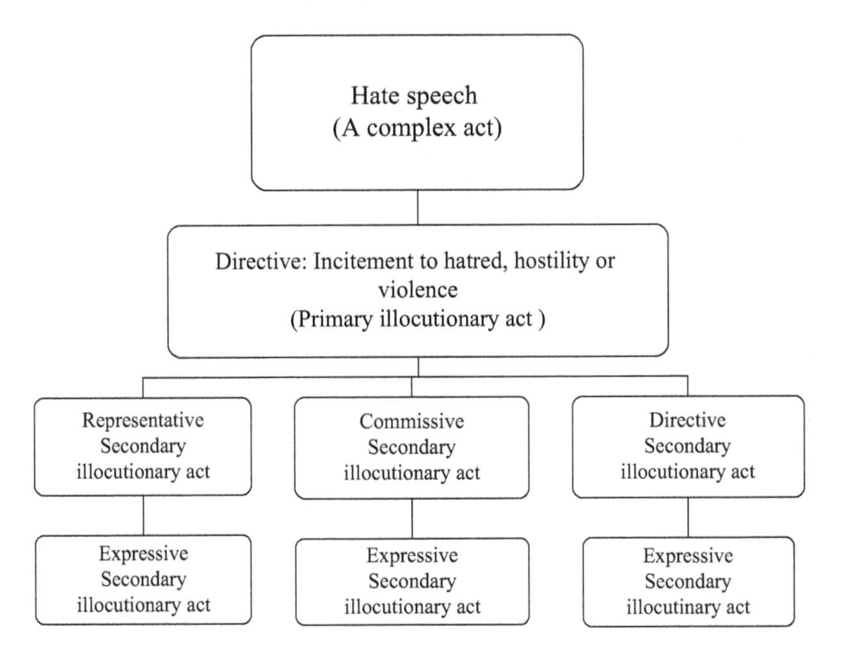

Figure 6.2: Hate speech as a complex act at a micro level.

interpret the illocutionary force of an utterance, though sometimes there may be a mismatch between the speaker's communicative intention and the hearer's interpretation of such meaning, as the next section shows.

7 Hate speech: The illocutionary force-perlocutionary link

Austin (1962) explained that any utterance involves the simultaneous performance of three acts:

1) The locutionary act: the act of producing a phonetically and grammatically correct utterance with semantic meaning (Searle 1980: 221–232).
2) The illocutionary act: the act containing the force or intention behind the words (Motsch 1980: 155–168; Searle & Vanderveken 1985; Alston 2000).
3) The perlocutionary act: the effect caused by the illocution on the thought and action of the addressee (cf. Davis 1980: 37–55).

When applying the above classification to hate speech, one finds that hate-advocating speakers (a) say something by using words (locutionary act), (b) do something by using words – e.g. disseminate, advocate or incite hatred, hostility or violence against the members of a group identifiable by legally-protected characteristics (illocutionary act) and (c) try to engage the audience in unlawful or violent acts against the targets (perlocutionary act). It is therefore arguable that incitement to hatred, hostility or violence, which is the main act of hate speech, is a prototypical example of a perlocutionary act because it is the effect of other acts associated with dissemination and advocacy, amongst others.

The illocutionary force-perlocutionary link is a core element in the legal approaches to hate speech. For instance, as shown in Chapter 2, US law may prohibit speech advocating the use of force or crime if the speech satisfies the Brandenburg test (Brandenburg v Ohio, 395 US 444, 1969): (1) The speech is directed to inciting or producing imminent lawless action (the illocutionary force of the utterance) and (2) the speech is likely to incite or produce such action (the perlocutionary act).

Another example of the link between the illocutionary force of an utterance and the perlocutionary act comes from international law: The Rabat Plan of Action (see Chapter 2) sets a high standard for limitations on freedom of expression, consisting of six criteria to determine where an expression creates such a danger of harm so as to justify prohibitions on expression: (1) the social and political context where the expression occurred, (2) the identity of the speaker – e.g. their status and influence over the audience, (3) the speaker's communicative intention, (4) the content and form of the expression, (5) the extent of the expression and (6) the likelihood and imminence of violence, discrimination, or hostility as a direct consequence of the expression. Criterion (3) points to the illocutionary force of the utterance, while criterion (6) points to the perlocutionary act.

Linguists and legal practitioners are concerned with interpreting the illocutionary force of an utterance unequivocally. In practice, this is not by any means an easy task. As shown in sections 4 and 5 of this chapter, both implicitness and, especially, indirectness, raise difficulties in interpreting the illocutionary force of an utterance. In theory, there seems to be a clear-cut correlation between a particular illocution and a certain grammatical mood – e.g. imperative, indicative or subjunctive and clause type – e.g. declarative, interrogative or exclamative. For example, an order (directive) like *Close the door, please!* is typically expressed with the imperative mood and an exclamative clause. It is a well-known fact that, in practice, the same order could be conveyed indirectly with the indicative mood and a declarative clause: *It is cold in here.* Since the form of an utterance does not give direct access to the speaker's communicative intention, the hearer has to infer the illocutionary force of the utterance by considering the context of situation in which the utterance is made (See Chapter 8).

Certain words such as racial epithets and slurs may also have more than one function. For instance, Home (2008) describes racial epithets as both insults (directives) and threats (commissives). Alternatively, Brown highlights the ambiguity of racial epithets by describing them as *polyfunctional constructs*:

> Using speech act analysis, we might conclude that the word *chink* or *nigger*, for instance, can be used potentially to perform [. . .] acts including but not limited to insulting, disparaging, degrading, humiliating, disheartening, harassing, persecuting, provoking, or inciting hatred, discrimination or violence (Brown 2017b: 32–33).

The additional problem is that the perlocution (Davis 1980: 37–55) caused by an illocution (Motsch 1980: 155–168) may not always issue from the speaker's communicative intention. A situation may arise in which the hearer does not secure uptake – that is, they do not correctly understand the illocutionary force of the utterance, the force which reflects the speaker's communicative intention. Uptake (Austin, 1962: 22, 116) is, in effect, one of the prerequisites for an illocutionary act to be felicitous (to be performed with success).

As far as uptake is concerned, there seem to be four scenarios open to linguistic analysis:

a) Speaker wants Hearer to engage in violent acts, and Hearer does secure uptake (Hearer engages in violent action).
b) Speaker wants Hearer to engage in violent acts, but Hearer does not secure uptake (Hearer does not engage in violent action).

c) Speaker does not want Hearer to engage in violent acts, but Hearer does not secure uptake (Hearer engages in violent action).
d) Speaker does not want Hearer to engage in violent acts, and Hearer secures uptake (Hearer does not engage in violent action).

Scenarios (a) and (d) depict situations with a perfect concurrence between the illocution and the perlocution. On the contrary, Scenarios (b) and (c) pose dilemmas for both linguists and legal practitioners. In both cases, the speaker may be equally indicted or even prosecuted because of the public dissemination of the hateful message (cf. Hancher (1980) and Guillén-Nieto (2020) on defamation).

As an illustration, I will now consider the illocution-perlocution link in Brandenburg v. Ohio (1969). The main purpose of the protest speech was to threaten the President, Congress and Supreme Court with violent retributive action if the state continued granting civil rights to African Americans and Jews. As explained earlier, the threat was conveyed implicitly through crafty linguistic choices, which show that Brandenburg was an experienced language user (see section 4 of this chapter):

> We're not a revenge organisation but if our President, our Congress, our Supreme Court, continues to suppress the white, Caucasian race, it's possible that there might have to be some revengeance taken. (Brandenburg v. Ohio, 395 US 4444 (1969), page U.S. 446).

From the perspective of Speech act theory, Brandenburg's threat (commissive) was felicitous because it satisfied the four felicity conditions: propositional content, preparatory condition, sincerity condition and essential condition. For the Ohio courts, the illocutionary force-perlocutionary link was clear in Brandenburg's utterance, as shown in the judgment. The Klan leader was charged under Ohio's Criminal Syndicalism Statute for his advocacy of violence as a means of political reform. However, the United States Supreme Court overturned Brandenburg's conviction. This controversial legal decision held that the local government could not constitutionally punish abstract advocacy of force or law violation unless it was directed to inciting or producing imminent lawless action and was likely to produce or incite such action (the imminence standard). It might be noted that the Supreme Court decision overlooked an essential aspect of the linguistic nature of threats that may contest the legal concept of imminent lawless action. Unlike a promise, when uttering a threat, the speaker does not need to intend to act but only to make the hearer fear that the speaker can carry out the threat. A threat may be considered an act whose perlocutionary effect is disembodied from its situation of reception and therefore unlikely to meet the imminence standard. Consequently, under US law, hate-advocating speakers might, then, exploit a threat's ambiguity to insulate them from prosecution.

The Case of Virginia v. Black (2003), also discussed in Chapter 3, offers another relevant example of the problems derived from the illocutionary force-perlocutionary link, though the criminal act in this case was not verbal. Rather, it was a symbolic expression performed through the physical act of burning a cross. For the Virginia court, it was clear that the defendants had burned the crosses to intimidate African Americans, violating a Virginia statute outlawing the burning of crosses to intimidate or cause fear. In the first case, a cross was lit on the property of an African American; in the second case, the cross was ignited at a Klan brotherhood rally organised in a private place. For the Virginia court, the physical act of burning a cross was equivalent to a true threat; hence, the desired perlocutionary effect was to cause fear, alarm or distress to African Americans. For the Supreme Court of Virginia, the physical act of burning a cross could not be taken as sufficient evidence of intimidation (*prima facie* evidence) – the perlocutionary effect was not so obvious – and it raised the decision to the Supreme Court. The majority opinion of the Supreme Court argued that the content of the *prima facie* evidence provision was unconstitutional because such provision carries the pragmatic presupposition that the burning of a cross must be automatically interpreted as intention to intimidate African Americans. The Supreme Court argued that burning a cross could also mean that a person is engaged in political speech. The reader will remember, from the discussion of the case in Chapter 3, that the Supreme Court struck down the Virginia statute to the extent that it considered burning crosses as *prima facie* evidence of an intention to intimidate (a true threat).

The case of A. v. The United Kingdom (2003), also discussed in Chapter 3, illustrates the unpredictable perlocutionary effects that a speaker's public speech may have on people and how these effects might arise irrespective of the speaker's communicative intention. As the reader will recall, during his speech in the House of Commons, an MP for the Bristol North-West constituency abused his absolute parliamentary privilege by referring to a young Black woman and her family as "neighbours from hell" and reporting on their antisocial behaviour offensively. In addition, excerpts of the MP's inflammatory speech and photographs of the young woman were released to the press. Even if the MP's communicative intention was not to defame and incite racial hatred towards a National citizen, he did: the young Black woman received hate mail addressed to her and threats; she was stopped in the street, spat at and called by strangers "the neighbour from hell"; the family had to be urgently rehoused, and the children moved to another school.

8 Conclusions

Ideally, creating a closed catalogue or inventory of the speech acts expressing hate speech would smooth the process of legal decision-making. In practice, such a task may be futile for the reasons indicated in this chapter, such as: the difficulty of assigning a speech act category to an utterance, the overlap between certain speech act categories, the need to interpret an utterance's illocutionary force when this is conveyed through implicit performatives and indirect speech acts. Although the application of Speech act theory to the analysis of hate speech is challenging due to the latter's complexity as an empirical object of study, the chapter has demonstrated the useful insights that this theory, grounded in the philosophy of language, can still provide at both macro- and micro-levels of linguistic analysis.

At a macro-level, Speech act theory offers linguists the necessary tools to dissect the pragmatic structure of hate speech. I argued that hate speech might be composed of acts specifically designed to accomplish one global perlocutionary act: to engage the audience in unlawful acts against members of a group identifiable by legally-protected characteristics. I claimed that the sequence includes three main acts: the preparatory act (dissemination), the auxiliary act (advocacy) and the main act (incitement). The proposed sequence structure accounts for hate speech's linear and hierarchical organisation as a global perlocutionary speech act. Each of the main acts in the sequence may also be assisted by other minor acts in the hierarchy, each pointing to the global perlocutionary act.

At a micro level, Speech act theory explained hate speech as a complex act comprising a primary illocutionary act (directive) and, depending on each case, one or more secondary illocutionary acts – e.g. representative, directive, commissive and expressive. It is noteworthy that expressive speech acts, whose function is to fuel hate, are often embedded in the primary illocutionary act or in the secondary illocutionary acts. Apart from being a complex act, hate speech is often safeguarded by the protective mantel of implicitness and indirectness, shielding hate-advocating speakers from indictment and criminal prosecution.

The chapter also clarified that hate speech should not be only associated with the expression of hate. Far from it, hate speech can, in effect, be the expression of other negative emotions and motives, such as anxiety, anger, frustration, or fear of loss of power and social privileges. Alternately, hate speech might be better categorised as speech motivated by hatred or speech capable of arising emotions, feelings or attitudes of hatred in others. This new concept, which is not considered in any speech act taxonomy, invites revision of the definition of expressive speech acts to account for the complexity of actual language use in context.

Another interesting insight provided by Speech act theory concerns the specific nature of incitement as a directive speech act. Because the hearer is not obliged to carry out the requested action (Preparatory condition), the speaker must persuade the audience using rhetorical strategies and conveying the message implicitly and indirectly instead of explicitly and directly. Speech act theory also explains the linguistic status of threats as typical features of hate speech. A threat is a commissive that expresses the speaker's communicative intention to carry out some harmful action against the targets. The speaker's purpose is to frighten people into believing they will be seriously harmed sometime in the future. Since the perlocutionary effect of a threat is delayed to some indefinite time in the future, it is reasonable to argue that the concept of a threat contests the imminence standard, especially in US law.

Lastly, the chapter also draws attention to the difficulties related to the illocutionary force-perlocutionary link, especially the likely mismatch between an utterance's illocutionary force, which reflects the speaker's communicative intention and its perlocutionary effect. The unpredictability of such a mismatch serves arguments that favour the hate-advocating speaker's civil liability when publicly communicating fighting words.

7 (Im)politeness theory

1 Introduction

In the early 1970s, Lakoff (1973) pioneered work on *politeness*, but, admittedly, it was Brown and Levinson's (1987 [1978]) ground-breaking formulation that laid the ground for research into politeness from a socio-pragmatic perspective. From Brown and Levinson's seminal work, and the work of others (Leech 1983) based on theirs to a greater or lesser extent, it has become clear in linguistic theory that people do not only speak to transfer information or do things with words but also to achieve a social goal: to relate to other people in a given socio-cultural context. An extreme example of how language can be used to fustigate or assault other people is hate speech. According to the Report of the United Nations High Commissioner for Human Rights on the Expert Workshops on the Prohibition of Incitement to National, Racial or Religious hatred – The Rabat Plan (2013) – hate speech requires the activation of a triangular relationship between the subject (the hate-advocating speaker), the audience (the in-group) and the object (the target), as depicted in Figure 7.1.

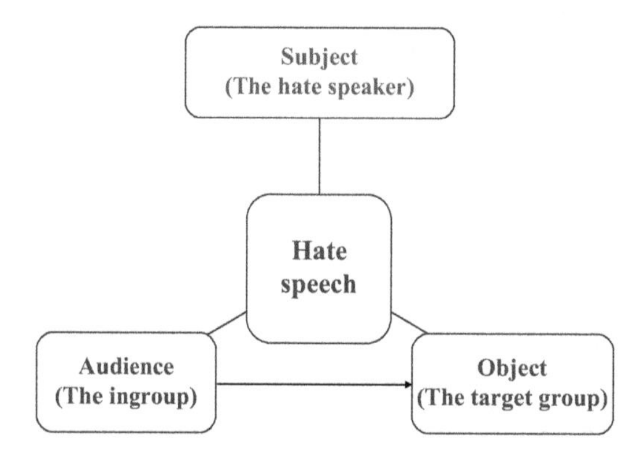

Figure 7.1: The participants in hate speech.

(Im)politeness theory offers an ideal framework for analysing hate speech, as it is specifically designed to harm the dignity and public image of the members of a group of people identifiable by legally-protected characteristics. This chapter points to the insights (Im)politeness theory can provide in the analysis of hate speech. To understand hate-advocating speakers' impolite behaviour it is first necessary to analyse their intentional deviation from polite behaviour. At the heart of (Im)politeness theory is this essential idea: polite behaviour entails

https://doi.org/10.1515/9783110672619-007

recognising that the addressee is a person with face wants like the speaker, while impolite behaviour does not entail such recognition. In the following sections, while I point to the main approaches taken by in (im)politeness theory, I also illustrate how the various theoretical views can be applied to examples of actual court cases of hate speech.

2 Politeness theory

Politeness,[116] as a basic social problem, tries to explain "how humans relate to other individuals of the same species and respond to societal wants" (Guillén-Nieto 2008: 200). The most influential approaches to politeness are the *conversational-maxim* approach (Lakoff 1973; Leech 1983, 2014; Gu 1990, 1997) and the *face-saving* approach (Brown & Levinson 1987 [1978]), theories to which I refer in the following sections in detail.[117]

---

116 The term politeness, as it is used in linguistics, is ambiguous and confusing because it refers to two different orientations. On the one hand, the term refers to conventional courtesy, social etiquette and good manners. This orientation is known as first-order politeness or politeness as social adequacy. On the other hand, the term is associated with strategic language use for face-saving purposes. This linguistic orientation is known as second-order politeness or politeness as a conversational strategy (Guillén-Nieto 2008: 201).

117 Apart from the conversational-maxim and face-saving approaches to politeness, another six main approaches can be found in the literature. The *social-norm* approach is associated with the understanding of politeness as social courtesy (Jespersen 1965; Quirk et al. 1985). The *conversational-contract* approach explains politeness as "operating within the then-current terms and conditions of the Conversational Contract" (Fraser 1990: 233), whose dynamic nature must be constantly adjusted to the changing terms and conditions of the ongoing conversation. The *emotive* approach explains politeness through the notion of interpersonal supportiveness, which is applicable to language, paralanguage and kinesics (Arndt & Janney 1979; 1983; 1985). The *relational* approach defines politeness as the work done by social actors who constantly negotiate their positions and, therefore, is part of a wider communicative experience (Watts 2003; Locher & Bousfield 2008). Besides, the idea of relational work offers an all-embracing perspective, covering not only politeness and impoliteness but also other kinds of relational practice, such as political behaviour. The *frame* approach defines politeness in terms of culture-specific ready-made patterns or frames acquired from the speaker's interactional experience with other people, stored in memory and easily retrievable in a given context (Terkourafi 1999: 107). The *rapport management* approach defines politeness as the management of interpersonal relations, which has two components: (1) face management – quality face (self-esteem) and identity face (social prestige) and (2) management of sociality rights – equity rights (personal consideration from others) and association rights (respect or friendship) (Spencer-Oatey 2000). Spencer-Oatey further distinguished between face-threatening behaviour and rights-threatening behaviour. In a later publication, Spencer-Oatey (2008) identified three sensitive levels affecting face: (1) the

2.1 The conversational-maxim approach

Lakoff (1973: 296) was the first linguist to draw attention to politeness as one of the two rules of pragmatic competence:

1. Be clear.
2. Be polite.

Lakoff noted that sometimes the need for clarity – Grice's *Cooperative Principle* (CP) and associated maxims (Grice 1975; 1989) – may clash with the need for politeness, in which cases one of the two rules takes priority, depending on the communicative situation. Lakoff (1973: 298) proposed three rules of politeness:

R1. Don't impose.
R2. Give options.
R3. Make the addressee feel good – be friendly.

According to Lakoff, R3 takes precedence over the other rules, because it embodies the purpose of all the rules of politeness: "The ultimate effect is to make the addressee feel good – that is, it produces a sense of equality between Sp[118] and A" (Lakoff 1973: 301). R3 makes the addressee feel wanted, as a more active participant, and highlights the significance of solidarity in social interactions.

I will now apply Lakoff's view of politeness to the analysis of hate speech. In doing so, the first relevant observation is that hate speech collides with the first two rules of politeness (R1 and R2) because it is a *monologic*[119] type of discourse. This authoritative type of discourse is, according to Bakhtin (1986), usually associated with the transmission of fixed ideas related to status inequalities. I argue that hate-advocating speakers want to plant the seeds of their ideology into someone else's mind. Hence, the meaning of a hateful text is fixed and imposed on the audience. The third rule (R3) seems to open two different scenarios, depending on whether the recipient of the message is a member of the ingroup (the audience) or a member of the target group (the object). Hate-advocating speakers will try to comply with R3 when they address their audience (the ingroup): they will try to make them feel good and of equal status. By contrast, hate-advocating speakers will overtly break R3 regarding the target group (the object) because

personal level, (2) the relational level and (3) the group level. This distinction implies that it is possible to threaten a speaker's face by underrating or degrading their person, close circle of friends, group or nationality, as in the case of hate speech.

118 In Lakoff's quote, "Sp" stands for the speaker and "A" for the addressee.

119 By contrast, Bakhtin (1986) claimed that dialogic discourse connotes social relations of equal status, intellectual openness and possibilities for critique and creative thought.

hateful messages are designed to inflict harm and damage the public image of the targets. To put it in Lakoff's words:

> The addressee will assume that we are no longer in a state of camaraderie; he will have been made to feel bad, a violation of the rules of politeness, rather than merely feeling he's been left his options or has not been imposed on (Lakoff 1973: 301).

To illustrate hate-advocating speakers' deliberate breach of R3, let us take Example 7.1, which reproduces an excerpt from Terminiello v. Chicago (1949), already analysed in the preceding chapters.

Example 7.1

It is the same kind of tolerance if we said there was a bedbug in bed [. . .] or if we looked under the bed and found a snake and said, "I am going to be tolerant and leave the snake there." We will not be tolerant of that mob out there. We are not going to be tolerant any longer.

So, my friends, since we spent much time tonight trying to quiet the howling mob, I am going to bring my thoughts to a conclusion, and the conclusion is this [. . .] I speak of the Communistic Zionistic Jew, and those are not American Jews. We don't want them here; we want them to go back where they came from (Terminiello v. Chicago, 337 US 1 (1949), p. 337 US 21).

As Example 7.1 shows, Terminiello, the speaker, tried to be friendly with his audience. He explicitly addressed the ingroups as "my friends" and employed the inclusive first-person plural ("We"), which blurs the speaker-audience divide by creating a sense of commonality, camaraderie and agreement. The inclusive "we" also strategically diminishes the speaker's responsibility for the speech content because they are portrayed as collaborating with the audience. By contrast, Terminiello referred to the target group, Jews and communists, in a highly offensive way, using fighting words, such as "bedbug", "snake", "the howling mob", the "Communistic Zionistic Jew" and "Jew". In addition, he alienated the targets in his speech, using the exclusive third-person plural ("They", "them"), and demanded that they be placed outside public life ("We don't want them here; we want them to go back where they came from"). As a result, the public image of any member of the target group was degraded and their equal rights were denied.

In sum, Terminiello did not deviate from the politeness rules because he was not a competent speaker. He deliberately opted not to comply with the politeness rules in performing a speech act – incitement to violence – that is clearly in conflict with the social goal pursued by politeness rules.

The other exponent of the conversational-maxim view is Leech (1983), who formulated a *Politeness Principle* (PP), inspired by Grice's Cooperative

Principle (CP)[120] and by Speech act theory. According to Culpeper and Terkourafi (2017), the PP[121] has explanatory power because whereas the CP accounts for how people convey indirect meanings, the PP accounts for why people convey them. The PP reads: "Minimise (other things being equal) the expression of impolite beliefs [. . .] Maximise (other things being equal) the expression of polite beliefs" (Leech 1983: 81). In a later publication, Leech described the PP as a:

> constraint observed in human communicative behaviour, influencing us to avoid communicative discord or offence and maintain or enhance communicative concord or comity (Leech 2014: 87).

Therefore, the PP paves the way for the analysis of impolite behaviour – that is, when the speaker deviates from the PP and opts not to avoid communicative discord maximising the expression of impolite beliefs, as in the case of hate speech.

Furthermore, Leech offered an illocutionary function classification that can be useful for analysing hate speech. Specifically, according to Leech (1983: 104–105), illocutionary functions can be classified into four subtypes,[122] depending on how they relate to the social goal of maintaining or enhancing concord or comity. These subtypes are:

a) Competitive: the illocutionary goal competes with the social goal – e.g. beg, order and demand.
b) Convivial: the illocutionary goal coincides with the social goal – e.g. congratulate, greet and thank.
c) Collaborative: the illocutionary goal is indifferent to the social goal – e.g. assert, inform and instruct.
d) Conflictive: the illocutionary goal conflicts with the social goal – e.g. threaten, accuse and insult.

Leech's taxonomy of illocutionary functions provides interesting insights into hate speech. On the other hand, the competitive illocution – the illocution goal competes with the social goal – dominates the social relation between the

---

120 Leech (1983) maintained that, unlike Grice's CP, which describes universal norms of communicative behaviour, the domain of politeness maxims is culture-specific because their weight may vary considerably from one culture to another. For a detailed study of politeness in Europe, see Hickey & Stewart (2005). Gu (1990; 1997) also offered an interesting view of politeness in China based on the so-called *Balance Principle.*

121 The PP has been criticised for three main aspects: (1) its universal pretensions, (2) its Anglocentric view and (3) its lack of practicality because of the many associated maxims.

122 Far from being water-tight compartments, Leech's four types of illocutions should be understood along the continuum of a cost-benefit scale (Leech 1983: 107).

subject (the hate-advocating speaker) and the audience (the ingroup). In this respect, the main hate speech act is a directive: incitement to hatred, hostility or violence. On the other hand, the conflictive illocution – the illocution goal conflicts with the social goal – dominates the social relation between the subject (the hate-advocating speaker) and the object (the target group).

Leech also adopted the notion of *pragmatic strategy* for its focus "on a goal-oriented speech situation, in which S uses language in order to produce a particular effect in the mind of H" (Leech 1983: 15). Two types of politeness strategies operate on Leech's illocutionary function taxonomy: positive and negative. Negative politeness minimises the impoliteness of competitive illocutions, which involve a higher cost for the hearer. By contrast, positive politeness maximises the politeness of convivial illocutions, which imply a higher benefit for the hearer. Politeness is considered irrelevant in collaborative illocutions and is out of the question in conflictive illocutions because they are innately offensive. For example, to threaten, accuse or insult someone in a polite manner is a contradiction and the idea can only make sense if the speaker does so ironically (Leech 1983: 105). From Leech's politeness strategies taxonomy, one can infer that the hate-advocating speaker will use negative politeness to minimise the impoliteness of the competitive illocutions directed at the ingroup (the audience) but will not have this obligation with the target group (the object). The reason for this is simple: the conflictive illocutions directed at the members of a target group are inherently offensive. The reader will remember from Brandenburg v. Ohio (1969) that Brandenburg employed negative politeness as a strategy to minimise the inherent impoliteness of the threat (conflictive illocution) directed at US authorities (see Chapters 3 and 6). In light of Leech's PP, the threat in Brandenburg's speech is ironic and raises particularised politeness implicatures (Culpeper & Terkourafi 2017: 9–10) (see Chapter 8).

Leech (1983) also provided some useful tools for measuring the degree of tact appropriate to a given situation:

a) The cost-benefit scale estimates the cost or benefit of the proposed action to S[123] or H.
b) The optionality scale orders illocutions according to the amount of choice S allows to H.
c) The indirectness scale orders illocutions according to the length of the path connecting the illocutionary act to its illocutionary goal from S's point of view. The indirectness scale can also be formulated from H's perspective according to the length of the inferential path by which the force is derived from the utterance meaning.

123 S stands for Speaker and H for Hearer.

Leech also added Brown and Gilman's pragmatic scales to his own pragmatic scales repertory (1960): (a) the power scale, which measures the degree of authority of one participant over another on the vertical axis and (b) the solidarity factor, which measures the social distance between the participants. These scales are useful tools when analysing hateful messages. The hate-advocating speaker will normally have a high degree of authority over the audience – e.g. the hate-advocating speakers were a priest in Terminiello v. Chicago (1940), KKK leaders in Brandenburg v. Ohio (1969) and Virginia v. Black (2003) and a Parliamentary MP in A. v. The United Kingdom (2003). The hate-advocating speaker will try to reduce the social distance with the audience, while they will enhance such social distance with the object. Since the cost (the degree of imposition) will be high for the audience, the subject will try to minimise their benefits by employing positive and negative politeness strategies, including indirectness. However, hate speech weighs low on the option scale. As noted earlier, it is monologic (Bahktin 1986) and hence imposes ideological meanings on the audience, leaving no room for creative thinking or debate.

Leech's PP is broken down into six maxims recommending the expression of polite rather than impolite beliefs (Leech 1983: 131–139):

1) The *Tact Maxim*: Minimise the expression of beliefs that express or imply cost to others.
2) The *Generosity Maxim*: Minimise the expression of beliefs that express or imply benefit to self.
3) The *Approbation Maxim*: Minimise the expression of beliefs that express or imply dispraise of others.
4) The *Modesty Maxim*: Minimise the expression of beliefs that express or imply praise of self.
5) The *Agreement Maxim*: Minimise the expression of beliefs that express or imply disagreement between self and others.
6) The *Sympathy Maxim*: Minimise the expression of beliefs that express or imply antipathy between self and others.

Whereas maxims (1) to (4) deal with bipolar scales: the costbenefit and praise-dispraise scales, maxims (5) to (6) deal with unipolar scales: the scales of agreement and sympathy.

It is reasonable to argue that hate-advocating speakers will try to employ all maxims in addressing the audience because they need to maintain or enhance communicative concord or comity. By contrast, hate-advocating speakers will implement none of the maxims when referring to the object because they pursue the opposite purpose: to maintain and enhance communicative discord and rudeness. In Example 7.1, when Terminiello spoke to his audience, he employed the Agreement Maxim – Minimise the expression of beliefs that express or imply disagreement between self and others. Interestingly, Terminiello's demands on the audience were presented as community agreement: "We don't want them here; we want them to go back where they came from." By contrast, Terminiello maximised confrontation with the object, as he deliberately broke the six maxims, especially the Sympathy Maxim – Minimise the expression of beliefs that express or imply antipathy between self and others. Terminiello employed highly offensive terms to refer to the targets: "bedbugs", "howling mob", "snakes" and "Communistic Zionistic Jew".

In a later publication, Leech (2014) reorganised the PP as a *General Strategy of Politeness* (henceforth, GSP):

> In order to be polite, S expresses or implies meanings that associate a favourable value with what pertains to O or associates an unfavourable value with what pertains to S (S = self, speaker; O = typically the hearer, but could be a third party) (Leech 2014: 90).

From the formulation of the GSP, one can infer that the hate-advocating speaker expresses or implies meanings that associate a favourable value with what pertains to the audience. Conversely, the GSP is absent when the hate-advocating speaker expresses or implies meanings that associate an unfavourable value with what pertains to the object.

2.2 The face-saving approach

The concepts of *rationality* and *face* are at the core of the face-saving approach, represented by Brown and Levinson (1987 [1978]). Rationality refers to competent speakers as rational agents who select the linguistic means necessary to satisfy their communication goals in a given communicative situation. The concept of *face*, which Brown and Levinson borrowed from Goffman (1967),[124] relates to a person's public self-image. The participants in conversation must

124 Goffman borrowed the concept of face from the Chinese language. In this lingua-culture, face has two sides: (1) an innate, subjective, personal sense of dignity and (2) a transcendent sense of dignity in social contexts. While the former relates to the individual's social standards of morality and behaviour (*lian*), the latter relates to the individual's social reputation and prestige (*mianzi*). Goffman uses the term face to refer to a person's public self-image, setting aside the subjective, personal sense of dignity. Interestingly enough, the personal side of face was later retrieved by Spencer-Oatey (2000) when she made a distinction between quality face (self-esteem) and identity face (social prestige).

constantly respect face because they are aware of their mutual vulnerability – that is, they can easily lose face in social interactions. Respecting each other's face wants is in the interest of every participant in conversation. In Brown and Levinson's view, a person's face wants are of two kinds: (1) *negative face* and (2) *positive face* (Brown & Levinson 1987 [1978]: 162). Negative face can be described as the want to have one's actions unimpeded by others, while positive face can be described as the want to have one's goals thought of as desirable. Face saving describes the lengths to which individuals may go to preserve their public image against face-threatening acts (FTAs). Face-threatening acts run contrary to the face wants of the hearer and the speaker, resulting in loss of face. In sum, any rational competent speaker must find a reasonable balance between the illocutionary goal (rationality) and the social goal (face) to avoid undesirable conflictive situations.

Moving into the arena of hate speech, hate-advocating speakers seem to be aware of the ingroups' face wants but systematically neglect the targets' face wants. This assertion implies that hate-advocating speakers do not care about or are indifferent to the targets' face wants. Degrading the targets' public image is, in effect, a discursive practice that promotes their exclusion from education, social wealth and mainstream society (cf. Delgado 1982; Matsuda 1989; Parekh 2006; Delgado & Stefancic 2018 [2004]). Moreover, vilifying the target also prevents loss of face for the hate-advocating speaker. To put it simply, the speaker can save face if the FTA is not directed at a human being but, instead, at a despicable object, such as "bedbugs", "snakes", "howling mobs", "niggers" and "chinks", amongst other derogatory terms. This strategy would explain the positive synergy between hate speech and social discrimination practices, such as dehumanisation, demonisation, marginalisation and negative stereotyping (cf. Stollznow 2017).

Brown and Levinson (1987 [1978]) distinguished the kinds of face threatened. Generally speaking, some acts threaten *negative face want*, while others threaten *positive face want*.

Acts that threaten the negative face want are:

a) Acts that primarily threaten the addressee's negative face want by indicating that the speaker does not intend to avoid impeding the hearer's freedom of action or freedom from imposition – e.g. orders, requests, threats and warnings.
b) Acts that predicate some positive future act of the speaker towards the addressee, putting some pressure on the addressee to accept or reject them and possibly incur a debt – e.g. offers and promises.
c) Acts that predicate the speaker's possible motivation for harming the addressee or the addressee's goods, giving the addressee reason to think that they may have to protect the object of desire or give it to the speaker – e.g. expressions of strong negative emotion such as hatred or anger towards the addressee.

Acts that threaten positive face want are:

a) Acts that indicate that the speaker has a negative evaluation of some aspect of the addressee's positive face want – e.g. negative evaluations, disapproval, contempt, ridicule, complaints and accusations.
b) Acts that show that the speaker does not care about or is indifferent to the addressee's positive face want – e.g. expressions of violent emotions in which the speaker gives the addressee possible reason to fear them or to be embarrassed by them – e.g. expressing irreverence, mentioning taboo topics, bringing up bad news about the addressee or boasting about the speaker, raising dangerously emotional or divisive topics, or using offensive or embarrassing address terms.

It might be noted that some FTAs, such as threats or violent expressions of emotions, intrinsically threaten both the positive face and negative face wants of the speaker and addressee, respectively.

From the above FTA classification, one can envisage the types of negative and positive FTAs hate-advocating speakers are likely to perform. These can be allocated differently, depending on whether the audience or the object are the recipients of hate speech:

1) Audience (the ingroups): negative face, types (a) and (b).
2) Object (the targets): negative face, types (a) and (c); and positive face, types (a) and (b).

Hate-advocating speakers will therefore perform acts that threaten the positive face of any member of a legally-protected group because, as Spencer-Oatey (2008) argued, it is possible to threaten an individual's face by underrating or degrading their group or nationality. Nevertheless, the hate-advocating speaker will not threaten the positive face of any member of the ingroup because the FTA would raise a conflict between the illocutionary goal (incitement to hatred, hostility or violence towards the object) and the social goal (maintaining and enhancing communicative concord and comity with the audience).

Additionally, Brown and Levinson (1987 [1978]) proposed five super strategies related to the degree of face threat, each super strategy encompassing several other lower-order strategies. They used the term *output strategies* to denote "the final choice of linguistic means to realise the highest goals" (Brown & Levinson 1987 [1978]: 92). According to Brown and Levinson's model, a rational competent speaker will then select an appropriate super strategy to counterbalance the expected face threat. To begin with, any rational actor may choose between two options:

1) Do the FTA.

2) Don't do the FTA.

If the speaker chooses the first option, then another two options will be available:
1) On record: The FTA will convey one unambiguously attributable intention.
2) Off record: The FTA will convey more than one unambiguously attributable intention so that the speaker cannot be held to have committed themselves to one particular intent. Some linguistic realisations of off-record strategies include metaphor, irony, rhetorical questions and understatement.

The idea that politeness is linguistically realised as indirectness has been challenged by cross-cultural empirical work (Hickey & Stewart 2005), which demonstrated that indirectness might be, at least in some cultures, associated with social hypocrisy or even authoritative or manipulative speech.

If the speaker chooses to perform the FTA on record, then they may choose between two strategies:
1) Bald on record.
2) With redressive action.

If the speaker chooses the second option, they may then choose to protect either the positive or negative face wants of the addressee:
1) Positive politeness is oriented towards indicating that the speaker respects the addressee's positive face want – e.g. treating the addressee as a member of the ingroup, with the same rights and duties as the speaker, presupposing, raising or asserting common ground, seeking agreement, giving understanding or cooperation to the addressee.
2) Negative politeness is oriented towards indicating that the speaker respects the addressee's negative face want – e.g. apologies for interfering, hedging the illocutionary force of the FTA, or using impersonalisation, e.g. passive sentences or nominalisation that distance the speaker and the addressee from the FTA.

Brown and Levinson (1987 [1978]: 74) recognised three contextual variables that together set the background for the operation of the strategies mentioned above: D(istance), P(ower) and R(anking) (of imposition in the particular culture). They proposed an equation for assessing the weightiness of an FTA as a function of three variables: W = D (S, H) + P (S, H) + R. D stands for the social distance between peers; Power stands for the relative power of the speaker over the addressee. It might be noted that power is not just related to physical force or wealth, as it can also refer to the control attributed by belief systems – e.g. a priest will have power over his flock (see Terminiello v. Chicago (1949)) or a

Klan leader over his followers (see Brandenburg v. Ohio (1969)). R stands for the level of imposition the FTA represents for the participants in the speech event.

Brown and Levinson's set of strategies may be considered a useful tool to assess the socio-pragmatic behaviour of the hate-advocating speaker. When the FTA is directed at the audience (the ingroup), this will be performed on record with redressive action (positive and negative politeness). On the contrary, when the FTA is directed at the object (the target group), it will be performed bald on record. Depending on the case, the FTA may also be performed off record. With this strategy, the hate-advocating speaker may try to avoid indictment or criminal prosecution (the concept of indirectness is discussed in Chapters 6 and 8).

In the following, some illustrative examples will be analysed. To begin with, let us revisit Example 7.1. In Terminiello's speech, one can observe several negative FTAs: (a) acts that threaten the negative face want of the members of the audience and (b) acts that predicate the speaker's possible motivation for harming members of the target group or their properties. In the first case, Terminiello opted for an off-record strategy: the FTA conveyed more than one unambiguously attributable communicative intention and hence Terminiello could not be held to have committed himself to one particular communicative intention. From the legal proceeding, it is known that the Chicago court prosecuted Terminiello for having incited violence. The FTA was not performed explicitly and directly but, instead, implicitly and indirectly through expressive and representative speech acts (see Chapter 6 on indirectness). Terminiello also employed an on-record strategy, with redressive action (positive politeness) when directing the FTAs at his audience. This strategy was oriented towards indicating that the speaker respects the positive face of the members of the audience by treating them as ingroups ("my friends", "We"), presupposing common ground and seeking agreement ("We must not lock ourselves in an upper room for fear of the Jews", "We don't want them here; we want them to go back where they came from"). Terminiello did not need to work on negative politeness because he, who had the power of moral authority over his folk, spoke on behalf of the ingroup, supposedly defending their social, economic and political interests. Terminiello did not have to take care of the negative face want of the ingroups because he and the ingroup shared the same negative face wants.

By contrast, Terminiello performed negative FTAs bald on record when referring to the object (the target group). The choice of this strategy indicated that he did not intend to avoid impeding the targets' freedom of action or freedom from imposition. Terminiello explicitly and directly imposed the ingroup's demand that the targets be placed outside public life in America: "We don't want them here; we want them to go back where they came from".

Example 7.1 also included positive FTAs: (a) positive FTAs indicating that Terminiello had a negative evaluation – contempt – of the targets ("howling mob") and (b) acts showing that Terminiello did not care about or was indifferent to the positive face of any member of the target group. Specifically, Terminiello deliberately used highly offensive address terms – that is, "bedbugs", "snakes" and "Communistic Zionistic Jew".

Example 7.2 reproduces an excerpt from Brandenburg v. Ohio (1969). Despite the 20-year gap between Brandenburg's speech and Terminiello's speech, they are very similar in discourse structure and strategy. Both speeches are, in effect, surface discursive manifestations of white supremacism and, to some extent, interdiscursively and intertextually related. In both speeches, the rhetoric divides the world into good (Us) and evil (Them), Americans and non-Americans, by constructing positive self- and negative other-presentation. In both speeches, the topoi of history and the saviour serve to construct revisionist historical narratives (cf. Wodak 2021).

Example 7.2

How far is the nigger going to – yeah.
This is what we are going to do to the niggers.
A dirty nigger.
Send the Jews back to Israel.
Let's give them back to the dark garden.
Save America.
Let's go back to constitutional betterment.
Bury the niggers.
We intend to do our part.
Give us our state rights.
Freedom for the whites.
Nigger will have to fight for every inch he gets from now on (Brandenburg v. Ohio, 395 US 4444 (1969), p. 395 US 449).

As a Klan leader, Brandenburg had the power of authority over the Klan members participating in the rally. Since the rally was broadcast, it reached a wider audience. In his speech, Brandenburg performed various negative FTAs: (a) acts that primarily threaten the negative-face want of the ingroups and (b) acts that predicate the speaker's possible motivation for giving the targets reason to think that they may have to protect themselves from harm to themselves or their properties. Brandenburg chose an off-record strategy and, therefore, could not be held to have committed himself to one particular intent: (a) Brandenburg could be protesting, on behalf of the Klan, against the US government's concession of civil rights to African Americans and Jews, (b) Brandenburg could be inciting his audience to hatred, hostility or violence against Jews and communists and (c) Brandenburg could be threatening US authorities. Brandenburg also performed FTAs on-record with redressive action and positive politeness by treating his audience, and white Americans in general, as members of the ingroup ("We"), presupposing common ground ("Save America", "Let's go to constitutional betterment", "We intend to do our part", "Give us our state rights", "Freedom for the whites") and seeking agreement ("This is what we are going to do to the niggers"). The use of negative politeness oriented to show respect to the negative face want of the ingroups was redundant because Brandenburg spoke on behalf of the KKK, defending their social privileges and economic interests. Brandenburg was the ingroup's voice.

On the contrary, when Brandenburg referred to the target groups, he performed the negative FTAs bald on record. This strategic choice indicated that he did not intend to avoid impeding the targets' freedom of action or freedom from imposition. Brandenburg imposed the ingroup's demand that the targets be placed outside public life in America: "Send the Jews back to Israel", "Let's give them back to the dark garden", "Bury the niggers".

Example 7.2 also illustrates positive FTAs performed bald on record, which indicates that Brandenburg did not care about or was indifferent to the positive face of any member of the target groups. Specifically, Brandenburg referred to the targets with racial epithets – e.g. "nigger", "niggers", "dirty nigger", "Jews" – whose derogatory force is, according to Mirón and Inda (2000: 102), "racial shaming".

Example 7.3 offers an interesting manoeuvre in Brandenburg's speech at the point he threatened US authorities.

Example 7.3

This is an organiser's meeting. We have had quite a few members here today which are – we have hundreds, hundreds of members throughout the State of Ohio [. . .] We're not a revenge organisation, but if our President, our Congress, our Supreme Court, continues to suppress the white, Caucasian race, there might have to be some revengeance taken (Brandenburg v. Ohio, 395 US 4444 (1969), p. US 446).

A threat is a commissive speech act that intrinsically threatens both the speaker and addressee's negative and positive face wants. Therefore, a threat is both a positive and a negative FTA. Brandenburg performed the negative FTA on record with redressive action and negative politeness, indicating that the speaker tried to minimise the impact of the FTA on the negative face want of US authorities. Specifically, Brandenburg used impersonalisation mechanisms and a passive construction in the second part of the conditional expressing the consequence or result ("[. . .] it's possible that there might have to be some revengeance taken"). As noted earlier, using positive politeness when performing a threat is incongruent because of the innate impoliteness it bears as a conflictive illocutionary act (Leech 1983). Hence, in this communicative situation, one can infer that Brandenburg used impersonalisation to protect his positive face want. The result was an ambiguous statement that also provided him with a shield from criminal prosecution.

Example 7.4 reproduces racist remarks from the TV interview of the Greenjackets in Jersild v. Denmark (1994).

Example 7.4

[. . .] the Northern States wanted that the niggers should be free human beings, man, they are not human beings, they are animals. Just take a picture of a gorilla, man, and then look at a nigger, it's the same body structure and everything, man, flat forehead and all kinds of things. A nigger is not a human being, it's an animal, that goes for all the other foreign workers as well, Turks, Yugoslavs and whatever they are called [. . .] (Jersild v. Denmark. Application No. 15890/89. Judgment. Strasbourg. 23 September 1994, p. 9).

The above racist remarks illustrate two types of FTAs towards the positive face want of any member of the target groups – Black people and immigrants: (1) acts that indicate that the speaker has a negative evaluation of some aspect of the addressee's positive face – e.g. contempt and ridicule – and (2) acts that indicate that the speaker does not care about or is indifferent to the targets' positive face want – e.g. by using offensive address terms. The Greenjackets performed such acts bald on record, implying that they did not care about the positive face want of any member of the target groups.

Example 7.5 is taken from A. v. The United Kingdom (2003), discussed in Chapter 3.

Example 7.5

[. . .] Neighbours from hell [. . .]

As the reader will remember from Chapter 3, in a parliamentary speech, a British MP referred to a British Black woman and her family as "Neighbours from hell" and reported on their anti-social behaviour. The speech was later publicly disseminated through the local and national press. As a result, the Black woman became the target of abuse from other British citizens to the point that she and her family had to be moved to another house. The derogatory expression: "Neighbours from hell" represents an FTA towards the positive face of the target. The act indicates that the speaker negatively evaluates some aspect of the target's positive face. In his parliamentary speech, the MP performed such act bald on record, indicating that he did not care about or was indifferent to the positive face want of the Black woman, and damaged her positive face want by underrating and degrading her racial group.

Example 7.6 reproduces a message posted by a member of the Proud Boys, an American far-right hate group that promotes and engages in political violence in the United States (see Chapter 6), on 9 May 2021 on the online forum *Proud Boys Uncensored*. The post also included a video.

Example 7.6

The black parasite who stabbed 2 Asian women was released from a 25-year prison sentence. The system actively gets these savages back on the street as soon as possible. That way they can continue doing the work of menacing decent people. @The WesternChauvnist.[125]

The above text illustrates FTAs towards the positive face want of the targets (Black people): (a) acts that indicate that the hate-advocating speaker has a negative evaluation of some aspect of the addressee's positive face want and (b) acts that indicate that the speaker does not care about or is indifferent to the addressee's positive face want by bringing bad news about a member of the target group and using offensive terms of address. The hate-advocating speaker performed such FTAs bald on record, implying that he did not care about or was indifferent to the target's positive face want.

Despite the undeniable impact and influence of Brown and Levinson's face-saving approach, today there seems to be general agreement that what individuals do in verbal interaction is not limited to avoiding face-threat, as they may intentionally threaten their interlocutor's face. It might be noted that even though Lakoff, Leech and Brown and Levinson prepared the ground for antagonistic or confrontational communication, their approaches focused on polite linguistic behaviour, setting aside, to a more or lesser extent, impolite linguistic behaviour as an object of study. At this point, Leech's words come to the author's mind:

> Presumably, in socialisation, children learn to replace conflictive communication with other types (especially the competitive type), which is one good reason why conflictive illocutions tend, thankfully, to be rather marginal to human linguistic behaviour in normal circumstances (Leech 1983: 105).

At present, Leech's remark might sound naïve. We know by experience that conflictive illocutions are rather frequent in language use, hate speech being

125 https:///www.lawenforcementtoday.com/ca-stabbing-suspect-was-granted-early-releasethrough-diversion-program/

an extreme example. The term impoliteness was coined because of the need to reflect a wider range of linguistic behaviour. On both sides of the Atlantic, Culpeper and Kaul de Marlangeon pioneered, in parallel, research into the phenomenon of impoliteness in the 1990s. I will now consider how impoliteness can expand the understanding of the examples analysed in this chapter.

3 Impoliteness theory

Unlike politeness, which concerns the individual's use of strategic language for face-saving purposes (Lakoff 1973; Leech 1983; Brown & Levinson 1987 [1978]; Watts, 2003), impoliteness[126] refers to the individual's use of strategic language for face-threatening purposes. Culpeper (1996; 2005; 2008; 2011; 2012; Culpeper & Terkourafi 2017) and Kaul de Marlangeon (1993; 2005; 2008; 2012 (with Alba-Juez); 2014) pioneered research into the use of strategies oriented to *face attack* and aimed at social disruption. Taking Brown and Levinson's politeness strategies as a point of departure, Culpeper (1996: 356–357) proposed an impoliteness framework consisting of five strategies:

1) Bald on record impoliteness. The use of strategies with a clear intention of attacking face directly – e.g. insults.
2) Positive impoliteness. The use of strategies designed to damage the addressee's positive face want – e.g. ignoring, excluding or being unsympathetic by denying association or common ground with the other.
3) Negative impoliteness. The use of strategies designed to damage the addressee's negative face want – e.g. scorning or ridiculing an interlocutor.
4) Sarcasm or mock politeness. The use of politeness strategies that are insincere – e.g. a sarcastic use of honorifics.
5) Withheld politeness. The absence of politeness work where it would be expected.

In later work, Culpeper and Terkourafi (2017) took an interactional perspective, arguing that

126 Impoliteness has become a multidisciplinary field of research that has attracted the attention of specialists from varied disciplines: social psychologists who deal with aggressive behaviour; sociologists who explore the effects of social abuse; specialists in conflict studies who analyse interpersonal and social conflicts, and linguists who study impolite linguistic behaviour from a socio-pragmatic perspective.

> (im)politeness is an evaluation that the speaker and addressee mutually reach by sequentially interpreting each other's utterances as they occur against the background of their prior experience both as members of society at large and with each other (Culpeper & Terkourafi 2017: 10).

An unexpected choice will therefore invite interactional awkwardness and invoke inferences. Within this framework, politeness and impoliteness are found within the spectrum of relational work, where appropriate relational behaviour is unmarked and inappropriate relational behaviour is marked. Unlike the unmarked relational options that tend to display a neutral emotion, the positively and negatively marked options are likely to express positive or negative emotions.

In the examples analysed in the preceding section, the hate-advocating speaker's behaviour can fall within inappropriate relational behaviour and is negatively marked because it expresses negative emotion – e.g. fear, contempt or hatred towards the targets. The hate-advocating speaker mostly opts to be overtly impolite by performing the FTA directed at a target group bald on record, indicating a clear communicative intention of attacking face explicitly and directly. Kaul de Marlangeon (2005) referred to these acts as *fustigation impoliteness* because they express the highest impoliteness force along the impoliteness continuum. Apart from employing a bald on record strategy, the hate-advocating speaker may sometimes opt to perform the FTA on record with some degree of redress, attacking the targets' positive or negative face.

Impoliteness is negatively marked because it is associated with negative social emotions: it may be caused by fear, contempt, anger, frustration, hatred, or arouse such emotions in the addressee. Moving to the arena of hate speech, the hate-advocating speaker often evokes an imaginary social conflict to vilify and marginalise the target group (cf. Delgado 1982; Delgado & Stefancic 2018 [2004]). Impolite linguistic behaviour overtly indicates that the hate-advocating speaker does not consider the members of the target group as persons like themselves with the same face wants. The hate-advocating speaker chooses to be deliberately impolite with the targets, attacking their positive and negative face wants. The following section discusses when *face damage* may become an actionable offence, such as *group defamation* or *group libel* (cf. Waldron 2012).

4 Impoliteness: Offence and moral damage

As shown in Chapter 1, hate speech may involve several reputational attacks that amount to assaults upon the human dignity and the social equality of the members of a target group (Delgado 1982; Matsuda 1989; Moran 1994; Ward

1997; Parekh 2006; Waldron 2012; Benesh 2014), and thereby inflict injury on individuals belonging to such groups (Massey 1992). Apart from incitement to hatred, hostility or violence, hate speech carries with it several auxiliary criminal offences, such as insults, threats, defamation and attacks on honour and moral dignity. Waldron (2012) suggested that hate speech should be considered group defamation or group libel because it involves damage to the victims' public image on account of their membership of a group identified by legally-protected characteristics.

In what follows, I will try to demonstrate that the theory of impoliteness also provides an analytical framework within which one can describe and explain the negative social emotions associated with offence and moral damage.[127] The impoliteness framework comprises three steps: (1) appraisal of behaviour in context, (2) activation of impoliteness attitude schemata and (3) activation of impoliteness-related emotion schemata (Culpeper 2011). Applying this impoliteness analytical frame to hate speech, one can understand how the offence is produced, perceived and evaluated in a communicative situation. For instance, consider the following: A speaker disseminates hateful messages towards members of a group identified by legally-protected characteristics (appraisal of behaviour in context); the speaker is thought to have acted in an inappropriately and unfairly hurtful way (activation of impoliteness attitude schemata); the speaker causes an emotional reaction in the targets such as embarrassment, shame, pain or anger, as a result of the speaker's face-threatening behaviour (activation of impoliteness-related emotion schemata). The negative emotion evoked in the targets triggers the possibility of bringing a claim for compensation of damage of a moral nature, provided that the speaker's abusive behaviour is considered an actionable offence in the jurisdiction where the offence must be judged. It is reasonable therefore to argue that the targets of hate speech may construct offence through a bottom-up process consisting of the following steps: (1) appraisal of behaviour in context, (2) activation of impoliteness attitude schemata, (3) activation of impoliteness-related emotion schemata, (4) appraisal of actionable offence and (5) claim for compensation of damage of a moral nature (cf. Guillén-Nieto 2020: 8).

---

127 Basic emotions such as happiness or sadness require awareness of one's somatic state. By contrast, social emotions related to societal norms and rights, such as embarrassment and shame, require representing other people's mental states. For the notion of social emotion, see Shaver et al. (1987: 1061–1086); Haidt (2003: 852–870); Hareli and Parkinson (2008: 131–156); Thomson and Alba-Juez (2014); Dewaele (2015: 357–370); Foolen (2016: 241–256); and Reeck, Ames and Ochsner (2016: 47–63).

Impoliteness occurs, as explained by Culpeper, "when: (1) the speaker communicates a face attack intentionally, or (2) the hearer perceives and constructs behaviour as intentionally face-threatening, or a combination of (1) and (2)" (Culpeper 2011: 19). As the author argues in a former paper on defamation (Guillén-Nieto 2020: 9), one may foresee four case scenarios:

A
a) Speaker communicates face attack intentionally and
b) Hearer perceives or constructs the Speaker's behaviour as face-threatening.

B
a) Speaker communicates face attack unintentionally and
b) Hearer perceives or constructs Speaker's behaviour as face-threatening.

C
a) Speaker communicates face attack intentionally and
b) Hearer does not perceive or construct Speaker's behaviour as face-threatening.

D
a) Speaker communicates face attack unintentionally and
b) Hearer does not perceive or construct Speaker's behaviour as face-threatening.

According to defamation law, only case scenarios A and B may involve defamation because the two preconditions for the crime are met. In the case of scenario A, there must be an intent-based element to damage the addressee's face and the latter, in turn, must secure uptake of the intended act. But the question is: How does one know what the speaker might have intended? Culpeper and Terkourafi (2017) pointed to the difficulty of determining a speaker's communicative intention:

> intentions are not stable private acts that are easy to grasp, but rather malleable constructs formed in interaction and may even be revised in light of future events (Culpeper & Terkourafi 2017: 11).

According to Malle and Knobe (1997: 111–112), intentionality requires the presence of five components: (1) a desire for an outcome, (2) a belief that a particular action will lead to that outcome, (3) an intention to act, (4) the skill to act and (5) awareness of fulfilling the intention while performing the act. The case of ES v. Austria (2019) will help us illustrate case scenario (a). As the reader will remember from Chapter 3, the Vienna public prosecutor's office brought charges against a woman for inciting religious hatred on the internet. The content of the statements, which were found incriminating by the domestic court, had a defamatory tone directed at the Prophet Muhammad. The European Court of Human

Rights upheld the conviction. In this case, I may conclude that the speaker wanted to communicate a face-attack because the five components required for intentionality to manifest itself were present: (1) the speaker wanted to damage the public image of Muslims (desire), (2) the speaker thought that disseminating discreditable information through online seminars (belief) might accomplish the desire, (3) the speaker, therefore, decided to communicate such information (intention), (4) the speaker knew how to do so (skill) and (5) the speaker was aware of fulfilling her intention (awareness) while conveying the defamatory messages. The speaker accused[128] the Prophet Muhammad of having been a child abuser without supporting her accusation with empirical evidence, thereby constructing a faulty generalisation: *All Muslims are paedophiles.*

Case scenario B would involve defamation because the target secures uptake. For example, in A. v. The United Kingdom (2003), discussed in Chapter 3, an MP made some defamatory remarks in his parliamentary speech directed at a Black woman and her family. The remarks incited racial hatred against the woman's family who had to suffer the negative consequences. The victim interpreted the derogative expression "neighbours from hell" and other degrading statements in the MP's parliamentary speech as positive and negative FTAs because of their defamatory nature. Unintentional perlocutionary acts indicate that our utterances may have an effect independent of our communicative intention. In the case of unintentional defamation, the defendant may finally mitigate damages or escape liability by offering an apology to the target, which did not happen in the case of A. v. The United Kingdom (2003). Instead, the MP claimed parliamentary immunity[129] in his defence.

5 Conclusions

This chapter was devoted to the insights (im)politeness theory can provide for the analysis of hate speech. At the heart of the theory is the concept that polite behaviour entails recognising that the addressee is a person with face wants like those of the speaker, while impolite behaviour does not entail such recognition. The chapter demonstrated that hate speech involves an intentional deviation on the part of the hate-advocating speaker from polite behaviour. The hate-advocating speaker uses language strategically for face-threatening purposes. This type of linguistic

---

128 Tiersma (1987) explains defamation as an act of accusation.

129 For a discussion on hate speech and absolute parliamentary immunity, see Brown and Sinclair (2020).

behaviour can be categorised as inappropriate relational behaviour that is negatively marked because of its association with negative emotions, such as fear, contempt, anger and hatred. This socially reprehensible type of behaviour invites interactional awkwardness and invokes inferences (see Chapter 8) about the hate-advocating speaker's communicative intention.

Through the conversational-maxim approach to politeness, one can learn that hate-advocating speakers deliberately break the pragmatic competence rules, as defined by Lakoff (1973), because they want to communicate other inferential meanings (particularised politeness implicatures) in a given communicative situation. Hate-advocating speakers flout Lakoff's pragmatic competence rules or Leech's (1983) PP because they, as experienced speakers, know how to exploit the pragmatic rules to maintain or enhance communicative concord and comity with the ingroups or communicative discord and conflict with the targets. The chapter pointed to the presence of positive and negative politeness strategies and associated maxims to minimise the impoliteness of the competitive illocutionary acts directed at the ingroups, especially the directive speech act: incitement to hatred, hostility or violence. By contrast, such politeness strategies and associated maxims were absent in constructing the conflictive illocutionary acts directed at the targets, such as threatening, accusing and insulting, mainly because of their innate offensive nature. In the same vein, Brown and Levinson's (1987 [1978]) face-saving approach improved the understanding of the types of negative and positive FTAs hate-advocating speakers can perform. It was also made clear that when directing negative FTAs at the ingroups, the hate-advocating speaker will try to save face by employing an on-record strategy with redressive action. Specifically, positive politeness will be employed to maintain and enhance communicative concord and comity within the ingroup, while an off-record strategy will be used to minimise the effect of the negative FTAs directed at the ingroups – e.g. order, demand and incite. When performed indirectly, the FTA conveys more than one unambiguously attributable intention; hence the hate-advocating speaker cannot be held to have committed themselves to one particular intent. By contrast, when the hate-advocating speaker directs positive FTAs at the targets, they will opt to be overtly impolite by mostly using a bald on-record impoliteness strategy or an on-record impoliteness strategy with some degree of positive or negative impoliteness (Culpeper 2011). In this vein, hate speech was categorised as fustigation impoliteness by Kaul de Marlangeon (2005) because of its intended rudeness towards the targets. The hate-advocating speaker intentionally threatens the targets' positive face want to cause harm to their public image.

Apart from explaining the hate-advocating speaker's strategic face-threatening behaviour, impoliteness theory can also be used as a working tool for the evaluation of offence and moral damage at a group level: It is possible to threaten a speaker's face by underrating or degrading their group or nationality (cf. Spencer-Oatey 2008). The linguist can assist triers of fact by identifying language cues about the speaker's intention of communicating face attack, and analysing how the addressee perceives and constructs the speaker's linguistic behaviour as intentionally face-threatening.

8 Cognitive pragmatics

1 Introduction

As mentioned in the previous chapter, hate speech requires the activation of a triangular relationship between the subject (the hate-advocating speaker), the audience (the ingroup) and the object (the target group). The subjects from the ingroup and the target group belong to different *interpretive communities*. This fact implies that they have different ideas, beliefs, assumptions and attitudes. It is predictable, then, that the subjects would bring with them different contexts or sets of assumptions in interpreting the speaker's communicative intention. Chapters 2 and 3 explained that hate speech is actionable in many jurisdictions if it incites hatred, hostility or violence against members of a legally-protected group. Chapter 6 showed that the hate-advocating speaker increasingly tends to incite hatred, hostility or violence implicitly and indirectly. Indirectness creates an additional difficulty in legal interpretation. When performed indirectly, a speech act conveys more than one unambiguously attributable intention; hence, the speaker avoids committing themselves to a particular communicative intention, leaving a part of the responsibility for the meaning to the hearer. This chapter elaborates on the meaning and interpretation of hate speech from the perspective of cognitive pragmatics. Specifically, the discussion focuses on how the speaker communicates intention – in this case, incitement to hatred, hostility or violence – that is not explicitly stated, and how this intention is likely to be interpreted by subjects belonging to different interpretive communities. I try to demonstrate the extent to which cognitive pragmatics may, on the one hand, improve our understanding of hate speech and, on the other hand, point to relevant linguistic clues for the interpretation of the speaker's communicative intention in court cases associated with hate speech.

2 Implicatures: The bridge from what is said to what is meant but not overtly said

Grice's (1975) key insight into pragmatic meaning was the distinction between *natural meaning* (what words mean) and *non-natural meaning* (what we mean by our words).[130] This distinction is based on the idea that the meaning of a

130 Levinson (2000) identified three types of meaning: (1) *entailment* (context-free and non-inferential), (2) *utterance-type meaning* (context-free and inferential) and (3) *utterance-token meaning* (context-sensitive and inferential).

https://doi.org/10.1515/9783110672619-008

message remains incomplete or undetermined if one only considers its natural meaning. Central to our discussion on the meaning of hate speech is Grice's *implicature*. As this chapter will show, this notion, which is at the heart of cognitive pragmatics, provides the bridge from what is said to what is meant but not overtly said. Via implicature, the speaker delivers information implicitly and indirectly, leaving it to the hearer to make assumptions about the speaker's intended meaning. Implicatures therefore depend on the hearer's capacity to draw inferences when an utterance's literal meaning is incongruent with the speaker's intended meaning. On this basis, it is reasonable to think that the speaker relies on the hearer's cognitive capacity and shared background knowledge to interpret the intended meaning.

2.1 Types of implicatures

Implicatures can be of two types: (1) *conventional implicatures* and (2) *conversational implicatures*. Whereas conventional implicatures are mainly derived from the natural (conventional) meaning of the words and expressions used, conversational implicatures are related to non-natural (non-conventional) meaning. Conversational implicatures are therefore inferences that have to be worked out through a combination of inferential processes in the context of the utterance. According to Grice (1975), conversational implicatures are processed against the background assumption that speakers and hearers observe the Cooperative Principle (CP) and its associated maxims – *Quantity*, *Quality*, *Relation* and *Manner*. A conversational implicature arises when a speaker intentionally flouts at least one of these maxims. In such cases, the hearer still assumes that the speaker is trying to be cooperative. Consequently, the hearer looks for the implicit meaning conveyed by making an inference that has its basis in the rational shared assumptions of the interlocutors (conversational implicature).

Conversational implicatures differ from conventional implicatures in the following distinctive properties (Grice 1975: 57–58):

a) *Cancelability* (*defeasibility*): The inference derived from the flouting of the maxims can be cancelled or defeated by adding some premises.
b) *Non-detachability*: Conversational implicatures are not attached to the linguistic form but to the content of what is said.
c) *Calculability*: Conversational implicatures can be inferred by using pragmatic principles and contextual knowledge.

d) *Non-conventionality*: Conversational implicatures are not generated from the conventional meaning of linguistic expressions.[131]

Grice (1975) also distinguished between generalised and particularised conversational implicatures. The former are inferences that refer to the non-explicit meaning that occurs by default in any context; the latter are inferences that are only derivable in a specific context.

We will now illustrate the notion of conversational implicature – the bridge between what is said and what is meant but not overtly said – by looking at two hypothetical communicative situations.

The first communicative situation is part of ordinary life. In this situation two friends (A and B) are planning to go to the beach. A says: *Shall we go to the beach, then?* B, looking at the black clouds in the sky, replies: *It is raining today*. If the inferences are drawn from the propositional meaning of the words used, A may conventionally imply that the weather is bad. However, Speaker A conveys an extra meaning that is not included in the utterance's propositional content and cannot be reduced to what B says. A's presupposed rationality and the CP allow A to infer that B is blatantly flouting the CP (*Maxim of Relation*) because he wants to convey another meaning that is not overtly said: *We cannot go to the beach*. A can infer B's intended meaning because A and B are rational speakers: B relies on A's cognitive capacity to infer the intended meaning by way of conversational implicature – in this particular case, a generalised conversational implicature, and A can interpret the apparent irrelevance of *It is raining today* as a way of being relevant and thereby meaning something else. Speaker A can infer B's intended meaning because of the pragmatic presupposition (*one cannot go to the beach if the weather is bad*) shared by both friends as rational speakers.

In the second communicative situation the analysis moves into the arena of hate speech, to consider an utterance made in a Klan rally (Brandenburg v. Ohio 1969), which was commented on in the preceding chapters. As is often the case with pragmatic notions, the notion of conversational implicature is described and explained through simple sentences or communication exchanges in fabricated contexts, like the first communicative situation above. The analysis of conversational implicatures in hate speech requires a discursive approach (De Beaugrande

131 Based on the works of Sadock (1978) and Horn (1991), Levinson (2000: 15) adds another two properties to Grice's list: (1) *reinforceability* and (2) *universality*. The first property refers to the fact that the speaker may reinforce the implied meaning by making it explicit. The second property refers to the universality of standard implicatures – that is, a given utterance should carry the same standard implicatures in any language into which it is translated.

2011; Bublitz 2011; Östman & Virtanen 2011; Sarangi 2011; Zienkowski, Östman & Verschueren 2011). For research purposes, I will illustrate the notion of conversational implicature through an exclamation Brandenburg uttered at the beginning of one of the speech fragments available in the court proceedings: "How far is the nigger going to – yeah". In making this utterance, the speaker conventionally implies that *Black people are going too far* (conventional implicature). The same utterance also implicitly conveys the speaker's intended meaning (conversational implicature). The problem is that, unlike the first communicative situation in which A and B are friends sharing the same background knowledge, in this situation, the speaker's utterance may be received by subjects who belong to different interpretive communities. The ingroups – Klan members and advocates – share with the speaker the same racist ideology (white supremacy) and prejudice, probably caused by fear of competition for wealth, territory, influence and social change (cf. Machin & Mayr 2012). As a result, the ingroups may infer that to restore American life's social standards, they must stop Black people's social advancement by using violence (particularised conversational implicature). Other recipients of the hateful message, such as the overhearers, bystanders or eavesdroppers, who may agree or disagree with the Klan's racist views, may infer that the speaker is protesting against US authorities for giving African Americans civil rights (particularised conversational implicature). African American people, the target group, may infer damage to their public social image, intimidation and a threat of impending violence against them (particularised conversational implicature). However, the critical question is: How can one establish what is to be inferred? The answer to this question is problematic for both linguists and legal practitioners when evaluating court cases associated with hate speech (cf. Baider 2020; 2022; Becker 2020; 2021). In the next section, I discuss how Sperber and Wilson's Relevance theory approaches this question by explaining the inference processes the hearer must go through to interpret the speaker's intended meaning.

3 The Principle of Relevance

Sperber and Wilson's (1995 [1986]) *Principle of Relevance* emanated from Grice's (1975; 1989) two central claims:

1) The claim that an essential feature of most human communication is the expression and recognition of intended meanings rather than a simple automated process of coding and decoding messages.
2) The claim that flouting the CP and associated maxims creates expectations that guide the hearer towards the speaker's intended meaning.

Whereas Sperber and Wilson fully supported the first claim, it is well known that they contended that the Principle of Relevance,[132] emanating from the Maxim of Relation, is, by itself, sufficient for inferring the speaker's intended meaning. In essence, the Principle of Relevance claims that: "Every act of ostensive communication communicates a presumption of its optimal relevance" (Sperber & Wilson 1995 [1986]: 158). The Principle of Relevance implies that conversation is always based on the assumption that there is an intention to communicate. Therefore, any act of communication automatically displays a presumption of relevance. Sperber and Wilson (1995 [1986]: 158) spelled out the presumption of optimal relevance as follows:

a) The set of assumptions that the communicator intends to manifest to the hearer is relevant enough to make it worth the hearer's while to process the ostensive stimulus.
b) The ostensive stimulus is the most relevant one the communicator could have used to communicate.

The Principle of Relevance then applies to both meaning production and meaning interpretation. On the one hand, the speaker will have to select the most relevant ostensive stimuli to attract the hearer's attention to their intended meaning.[133] To infer the speaker's intended meaning, the hearer will have to choose the relevant context:[134] the set of assumptions that produce the maximum benefit (greatest contextual effects) with the minimum cost (lowest cognitive processing effort). I concur with Escandell-Vidal (1996: 120) that being relevant is not an intrinsic characteristic of utterances. Rather, it is a property that arises from the relationship between the utterance and the hearer with their particular set of assumptions in a given context. Consequently, what may be relevant for one person may not be relevant for somebody else.

---

132 Sperber and Wilson (1995 [1986]) use the term *relevance* as a technical term to describe the degree of cognitive effort required for a hearer to infer the speaker's intended meaning.

133 The Principle of Relevance tries to flesh out the hearer's steps in constructing inferences to discover the speaker's communicative intention and an interpretation consistent with the Principle of Relevance. One of the criticisms the theory has received points out that it does not explain the steps the speaker must take to produce a relevant utterance (cf. Escandell-Vidal 1996: 130–133).

134 In Relevance theory, the notion of context refers to the set of assumptions or mental representations of the real world used in interpreting an utterance. The context is not given in advance. On the contrary, the context is chosen by the hearer of an utterance at any given moment. The hearer searches amongst their total set of assumptions for those that lead them to the most relevant possible interpretation.

3.1 Ostensive-inferential communication

At the core of Relevance theory is the notion of *ostensive-inferential communication* that Sperber and Wilson defined as follows:[135]

> the communicator produces a stimulus which makes it mutually manifest to communicator and audience that the communicator intends, by means of this stimulus, to make manifest or more manifest to the audience a set of assumptions (Sperber & Wilson 1995 [1986]: 63).

The ostensive stimuli the speaker can select in the speech production process are both linguistic[136] and non-linguistic – e.g. paralinguistic and kinetic features and visual elements. The hearer should recognise the ostensive stimuli pointing to the communicative intention behind the utterance – that is, what is meant but is not overtly said. Interpretation, like other cognitive processes, works through heuristic reasoning. This type of reasoning is not entirely falsifiable,[137] as the hearer may be wrong in the inference process. In such a case, neither the presumption of relevance nor the validity of the reasoning implied by the hearer is flawed. On the contrary, the problem is that the inference is made from different or wrong assumptions. Therefore, the implicated conclusion the hearer reaches is not the speaker's intended one. Since it is not possible to be absolutely certain about the speaker's communicative intention, the hearer can only set up a hypothesis that satisfies the presumption of relevance conveyed by the utterance.

---

135 Human communication consists of two layers: (1) codification-decodification communication and (2) ostensive-inferential communication. The first layer describes how thoughts are coded in utterances and, subsequently, utterances are decoded in thoughts. The correspondence between signs and messages is pre-established and conventional in this first layer. The second layer describes the speaker using ostensive stimuli to attract the hearer's attention towards the communicative intention behind the utterance. In this second layer, the correspondence between actions and intentions is non-conventional; hence, the hearer must always interpret it in a given communicative situation.

136 Note that the encoded message can also function as an ostensive stimulus. For example, the utterance *It's raining cats and dogs* encodes a message other than the one intended to be communicated: *It's raining a lot*. The utterance *It's raining cats and dogs* points to another reality that invites the hearer to make the inference that allows them to recover the speaker's intended meaning.

137 Mey and Talbot (1988) argue that it is not legitimate to construct an explanation of communication connecting actions with intentions because one cannot have direct access to the speaker's cognitive structures and mental representations. Consequently, the connection between actions and intentions is presented as a reasonable but non-falsifiable hypothesis. In other words, one can only assume that since actions reflect intentions, their study can lead one to them.

Wilson and Sperber (2007: 615) explained that the general task of constructing a hypothesis can be broken down into several subtasks, not necessarily carried out in sequence, each involving a specific inference process:

1) Constructing an appropriate hypothesis about explicit content or *explicature*[138] via decoding, disambiguation, reference resolution and other pragmatic enrichment processes.[139]
2) Constructing an appropriate hypothesis about the intended contextual assumptions (*implicated premises*).
3) Constructing an appropriate hypothesis about the intended contextual implications (*implicated conclusions*).

To illustrate the above deductive process, let us take a hypothetical example in which there are two friends, A and B. A asks B what time it is.

Example 8.1
A: What is the time?
B: The bus is coming.

A's deductive process can be represented as follows:

Explicature
a. B can see the bus approaching the bus stop.

Implicated premise
a. We both know that the bus always arrives at the bus stop at 10 o'clock.

Implicated conclusion
a. It is almost 10 o'clock.

The interpretation of B's utterance requires three different deductive steps: A completes the semantic representation of B's reply by constructing the explicature or propositional meaning of *The bus is coming* via decoding and reference resolution. A must supply some premises, especially the reasoning that links

---

138 The notion of explicature refers to an enrichment of the explicitly encoded information in an utterance into a fully elaborated propositional form.

139 It consists of fleshing out the skeletal propositions needed to complete an utterance. Two main kinds of enrichment are (1) the recovery of missing elements in the case of ellipsis and (2) the resolution of semantic incompleteness.

the meaning in the question with the meaning provided by the answer. Implicated premises must be retrieved or constructed by developing assumption schemas[140] from memory. In this case, the implicated premise can be deduced from the shared knowledge of the participants in conversation: *We both know that the bus always arrives at the bus stop at 10 o'clock.* The implicated conclusion is deduced from the explicature and the context: *It is almost 10 o'clock.* Of course, in Example 8.1, B could have replied explicitly: *It is almost 10 o'clock.* Still, the explicature would not have been as relevant as the particularised conversational implicature in which two assumptions are made manifest by processing a single utterance: *The bus always arrives at the bus stop at 10 o'clock* and *It is almost 10 o'clock.*

3.2 Contextual effects

The relevance of an utterance is reflected in the *contextual effects* – that is, the extent of the impact the utterance has on the information currently available to the hearer. Wilson and Sperber (2007: 609) formulate the relevance of information to a subject in these words:

a) Other things being equal, the greater the positive cognitive effects achieved by processing an input, the greater the relevance of the input to the individual at that time.
b) Other things being equal, the greater the processing effort expected, the lower the relevance of the input to the individual at that time.

One can then conclude that the less cognitive effort it takes to recover information, the greater its relevance.

Additionally, Cruse (2004: 383) pointed out that there are four kinds of contextual effects: (1) adding new information, (2) strengthening old information, (3) weakening old information and (4) cancelling old information.

140 The relative importance of the assumptions is essential in drawing inferences. A strong assumption will produce a strong inference, while a weak assumption will produce an invalid inference. According to Sperber and Wilson (1995 [1986]), an assumption's greater or lesser weight depends on how the subject acquires it. For example, when an assumption results from a subject's direct experience, its weight is greater than if the subject had learned the assumption from other people. The power or moral authority of the speaker also gives greater weight to the assumption than if it comes from a speaker with little or no power or moral authority.

3.3 Salience

Salience is a semiotic notion referring to the relative prominence of signs. The salience of a particular sign in a given context assists subjects in ranking large amounts of information by importance, hence paying attention to the most relevant information. According to Kecskes:

> In pragmatics, when we speak about salient information, we mean given information that the speaker assumes to be in a central place in the hearer's consciousness when the speaker produces the utterance (Kecskes 2014: 176).

The most relevant interpretation of an utterance, which is normally the most salient, may sometimes be an explicature, sometimes an implicature. When there is more than one possible explicature or implicature, salient information helps the hearer choose the most relevant one – that is, the one with the greatest contextual effects and lowest cognitive processing effort. It might be noted that salience can also be perceptual and apply to visual ostensive stimuli in conversation. Although salient meaning is easy to access for any competent language speaker, sometimes it might not be easy to decide the most salient inference to be drawn from an utterance. As Leech pointed out:

> The indeterminacy of conversational utterances [. . .] shows itself in the negotiability of pragmatic factors; that is, by leaving the force unclear, S may leave H the opportunity to choose between one force and another and thus leaves a part of the responsibility of the meaning to H. For instance, *If I were you, I'd leave town straight away* can be interpreted according to the context as a piece of advice, a warning, or a threat, and action it as such; but S will always be able to claim that it was a piece of advice [. . .] (Leech 1983: 23–24).

Malicious speakers may then exploit the indeterminacy of conversational utterances to express their intended communicative purpose under cover. For this reason, inferences must always be worked out in the context of situation in which the utterance is expressed.

4 The Principle of Relevance applied to court cases associated with hate speech

This section revisits some of the court cases associated with hate speech discussed in the preceding chapters. All the cases selected demonstrate how legal reasoning and decisions about the speaker's intended meaning can differ dramatically from one court to another. The problem is, as noted, that interpretation is not entirely falsifiable, because it works through heuristic reasoning. Since the hearer's inference can be made from different or mistaken assumptions, they can reach an implicated conclusion that, though relevant, is incongruent with the speaker's intended meaning. The hearer can only set up a hypothesis about the speaker's intended meaning that satisfies the presumption of relevance conveyed by the utterance. I argue that the more salient ostensive stimuli the speaker deliberately uses in the speech production, the more certain the hearer's hypothesis on the speaker's intended meaning can be in the interpretive process. Rational behaviour assumes that we do things to achieve a particular purpose. It might therefore be argued that the speaker selects the most salient ostensive stimuli that they think are in the hearer's mind to attract their attention toward the utterance's communicative intention. Since it is impossible for anybody, including linguists, to peer into the speaker's mind to discover their communicative intention, one must examine every act of ostensive communication displayed in the speech act to deduce the speaker's intended meaning.

Applying the Principle of Relevance to hate speech requires an analysis beyond the sentence level; linguists must overcome the difficulty of analysing large, complex discourse units instead of the simple sentences or conversational exchanges that are typical explanations of the theory. Interpreting live data is, in effect, more demanding than interpreting fabricated data. For instance, in actual conversations the hearer must process and interpret a continuous and rapid flow of information. Besides, the hearer must engage in multiple concurrent inferencing processes.

4.1 Ostension coded in language

This section uses the conclusion of Terminiello's speech, which is available in the court proceedings of Terminiello v. Chicago (1949), to illustrate ostention coded in language. Drawing on the Principle of Relevance, the author assumes that the conclusion of Terminiello's speech is an act of ostensive communication that automatically displays a presumption of optimal relevance. It must be noted that Terminiello's speech was addressed to an audience with whom the speaker shared the same prejudice and intolerance towards Jews and communists. The speaker and the audience together were part of the same interpretive community. I argue that the linguistic ostensive stimuli Terminiello selected were salient to the audience and therefore the most relevant ones he could use to make his communicative intention manifest.

The first ostensive stimulus that stands out in Terminiello's speech conclusion comes from the expression of fighting words in the form of dehumanising metaphors – e.g. "bedbug", "snake"- and slur terms – e.g. "slimy scum", "dirty kikes", "Communistic Zionistic Jew". According to the legal definition, these terms can be considered fighting words because (a) they are directed at the person of the outgroups, (b) by their very utterance, they inflict injury and tend to provoke a violent reaction and (c) they play no role in the expression of ideas (cf. Purtell v. Mason (7thCir. 2008)).[141]

Apart from fighting words, Terminiello's speech conclusion included other salient linguistic ostensive stimuli for his racist audience. On the one hand, the opposition between the personal pronouns Us and Them signals social polarisation between the ingroup (white Americans) and the outgroup (Jews and communists). White Americans are presented as victims, while the targets are dehumanised and vilified (see van Dijk 1992 on blaming the victim and victim-perpetrator reversal). On the other hand, deontic modality is employed to indicate that the state of the world, which must be understood as the surrounding circumstances, does not meet certain social standards (laws) or personal standards (desires). The utterance containing the deontic modal indicates some action that would change the world so it can become closer to social and personal standards. Specifically, in Terminiello's speech conclusion, three subcategories of deontic modality emerge: (1) commissive modality expressing the speaker's commitment to do something ("We will not be tolerant [. . .]"), (2) directive modality expressing an obligation placed on the hearer ("We must [. . .]") and (3) volitive modality expressing the speaker's wishes ("We (don't) want [. . .]").

These ostensive stimuli make it mutually manifest to the speaker and intended audience that the former intends to communicate a set of assumptions (cf. Sperber & Wilson 1995 [1986]: 63). Although the hearer can never be absolutely sure about their constructed hypothesis, salience helps them choose the most relevant assumptions – that is, the ones producing the greatest contextual effects with the lowest cognitive processing effort. To fill the gap from what is said to what is meant but not overtly said (conversational implicature), the hearer can only construct a hypothesis about the speaker's intended meaning that satisfies the presumption of relevance conveyed. As noted earlier, the general task of constructing a hypothesis can be broken down into three subtasks. For the purposes of analysis, I refer here only to some of the hypothetical inferences the audience might make to find out the speaker's intended meaning. Specifically, the interpretation of Terminiello's speech conclusion requires three deductive steps:

1) The audience completes the semantic representation of Terminiello's speech conclusion by constructing a hypothesis about the explicatures or propositional meanings via decoding and reference resolution:

---

141 https://caselaw.findlaw.com/us-7th-circuit/1197679.html (accessed 21 May 2022).

Explicatures
a. *Jews and communists are compared to parasites* ("bedbugs") *and reptiles* ("snakes").
b. *We are not tolerant of parasites* ("bedbugs") *and reptiles* ("snakes") *because they are harmful and dangerous.*
c. *We will not be tolerant of Jews and communists.*
d. *We must not be fearful of Jews and communists.*
e. *We want Jews and communists to leave America.*

2) The audience constructs hypotheses about the implicated premises. These can be deduced from the white supremacist ideology shared by the speaker and thier intended audience that together form the ingroup:

Implicated premises
a. *Jews and communists are not like us.*
b. *Jews and communists are inferior to us.*
c. *Jews and communists are outsiders.*
d. *The massive arrival of Jews and communists is putting white American social standards under threat.*
e. *Jews and communists are stealing our jobs, properties and wealth.*

3) The audience constructs a hypothesis about the implicated conclusions. These are deduced from the explicatures and the context (set of assumptions) the audience selects for the interpretation of Terminiello's speech conclusion:

Implicated conclusions
a. *Jews and communists are like a plague infesting and harming America.*
b. *We hate Jews and communists.*
c. *We must fight off Jews and communists.*

In the second deductive step – the construction of the hypothesis of the implicated premises – the members of the audience or ingroups must select from among their most relevant ideas, beliefs, assumptions and attitudes to interact with the new assumptions provided by the propositional meaning of the explicatures. At this step of the deductive process, one can expect different implicated premises from one interpretive community to another, especially from the ingroup and the outgroup, and therefore one can also expect differing, though still relevant, implicated conclusions. As earlier noted, relevance is not an intrinsic property of an utterance but a quality that depends on the set of assumptions brought by the hearer to interpret the speaker's intended meaning. From this hypothetical deductive process, the audience would reach the implicated conclusions that satisfy the presumption of relevance conveyed by Terminiello's words. His speech conclusion is relevant for the audience because it produces the greatest contextual effects with the lowest cognitive processing effort. The contextual effects add new information but mainly strengthen the audience's negative assumptions (prejudice and intolerance) towards Jews and communists.

In this case, the linguistic clues provided by the Principle of Relevance support the hypothesis that Terminiello intentionally incited the audience to hatred and imminent violence against the targets protesting outside the auditorium where he was giving his inflammatory speech. As the reader will remember from the discussion in Chapter 3, Terminiello was arrested for riotous speech and found guilty by the court for violating Chicago's breach of the peace ordinance. Subsequently, the majority opinion of the Supreme Court overturned Terminiello's conviction, protecting his right to freedom of expression under the First Amendment to the Constitution of the United States.

4.2 Multimodal ostension

Many court cases associated with hate speech involve more than one mode of ostensive communication. Such is the case of Brandenburg v. Ohio (1969), to which I have already referred. The evidence the courts of justice had to examine included spoken language and television footage. I assume that Brandenburg selected the most salient ostensive stimuli for his audience – Klan members and their advocates – to attract their attention towards his communicative intention. In what follows, the reader will see that the selected ostensive stimuli were, in effect, very similar to Terminiello's. To begin with, Brandenburg employed fighting words in the form of racial epithets – e.g. "nigger", "Jews" and "dirty nigger". According to Homs (2008: 426–430), the racial epithet "nigger" is considered extremely derogatory towards African Americans because it has the power of invoking white supremacy and the active, pernicious and wide-ranging racist institutions that support it. Therefore, on no grounds can the derogatory content of racial epithets be considered accidental in Brandenburg's speech. In addition, the mentioned fighting words were reinforced by overlexicalisation: The over-presence of "nigger" and "Jew" contributed to the performative act of shaming the targets. As Mirón and Inda claimed, these racial epithets are:

> not so much a constative utterance, a statement of fact, as a performative through which a racial subject is produced and shamed. [. . .] The shaming of a subject only acquires a naturalised effect through repetition (Mirón & Inda 2000: 102).

Brandenburg also resorted to other linguistic ostensive stimuli. The opposition between the personal pronouns Us and Them served to express social polarisation between the ingroup – the Klan (the saviour) and white Americans (the victims) – and the targets (African Americans and Jews). Brandenburg also employed two forms of deontic modality: (1) commissive modality expressing the speaker's commitment to do something: "This is what we are going to do to the niggers", "Send the Jews back to Israel", "Let's give them back to the dark garden", "Bury the niggers", "We intend to do our part", "Save America", "Nigger will have to fight for every inch he gets from now on"; and (2) directive modality expressing an obligation placed on US authorities: "Give us our state rights", "Freedom for the whites".

Since the protest speech was broadcast, the visual ostensive stimuli selected by Brandenburg were foregrounded. For instance, viewers of the broadcast could see Klan insignia: the Klan leader wore a red hood and robe signalling leadership, while the Klan members wore white hoods and robes. They all gathered around a burning cross (the relevance of cross burning as a hate symbol is discussed in section 4.2.1 of this chapter). Other salient visual ostensive stimuli were a Bible, symbolising moral authority, and firearms and ammunition, representing power and potential violent action. The interpretation of Brandenburg's words requires three different deductive steps:

1) The audience completes the semantic interpretation of Brandenburg's words by constructing the explicatures or propositional meanings via decoding and reference resolution:

   Explicatures
   a. *Niggers and Jews are going too far.*
   b. *The Klan wants Jews to go back to Israel.*
   c. *The Klan wants niggers to disappear.*
   d. *The Klan demands state rights and freedom for white Americans from US authorities.*
   e. *The Klan has many followers all over the US.*
   f. *The Klan has weapons.*

2) The audience constructs a hypothesis about the implicated premises. These can be deduced from the white supremacist ideas, beliefs, assumptions, values and attitudes shared by speaker and intended audience (the ingroup):

   Implicated premises
   a. *African Americans and Jews are not like white Americans.*
   b. *African Americans and Jews are inferior to white Americans.*
   c. *African Americans and Jews are outsiders in America.*

d. *African Americans and Jews must not have the same state rights white Americans have.*
e. *Because of the civil rights movement, African Americans and Jews have the same state rights white Americans have.*
f. *The more state rights African Americans and Jews have, the more powerful they are.*
g. *The more powerful African Americans and Jews are, the less powerful white Americans are.*
h. *The Klan has moral authority to defend America.*
i. *The Klan is a powerful organisation.*

3) The audience constructs hypotheses about the implicated conclusions. These are deduced from the explicatures and the context (set of assumptions) the intended audience selects for the interpretation of the message:

   Implicated conclusions
   a. *We must restore America's social standards.*
   b. *We must stop the social advancement of African Americans and Jews in America.*
   c. *We must take violent action against African Americans and Jews.*
   d. *We are ready to fight US authorities unless they suppress state rights to African Americans and Jews.*

As shown in Terminiello v. Chicago (1969), the construction of the implicated premises is crucial in the deductive process for reaching the implicated conclusions. At this stage, the ingroup and the outgroup will construct different hypotheses based on their respective assumptions and infer different intended meanings. To interpret the conversational implicatures of Brandenburg's words, the audience – Klan members and their advocates – would recur to their white supremacist views. Brandenburg's speech is relevant to the intended audience audience because it produces the greatest contextual effects with the lowest cognitive processing effort. The contextual effects add information, but mostly they reinforce the ingroups' negative beliefs and assumptions (prejudice and intolerance) towards the targets. The reader will recall from the discussion in Chapter 3 that Brandenburg was charged under Ohio's Criminal Syndicalism Statute for advocacy of violence or crime as a means of political reform. The Supreme Court overturned his conviction on the grounds that it was unconstitutional to punish abstract advocacy of force or law violation unless it is directed to inciting or producing imminent lawless action and is likely to produce or incite such action (the imminence standard). The linguistic cues provided by the Principle of Relevance would support the hypothesis that Brandenburg deliberately incited hatred and violence towards African Americans and Jews. One should also remember the socio-political context in the 1960s: the Klan's terrorist attacks against African Americans, Jews and civil rights defenders were shockingly frequent (see Chapter 4).

4.2.1 The relevance of hate symbols

Hate speech also includes the public display of hate symbols. In this case, the speaker selects symbols as salient visual ostensive stimuli to help the audience interpret the speaker's intended meaning. Three of the court cases discussed in Chapter 3 are related to the public display of a controversial symbol: National Socialist Party v. Skokie (1977), Virginia v. Black (2003) and Fáber v. Hungry (2012). In what follows, I will examine each of them in further detail.

At the core of the legal reasoning of National Socialist Party v. Skokie (1977) was the question of whether or not a hate symbol is sufficient evidence of intimidation (*prima facie* evidence). The reader will remember from the case records that the village of Skokie, where many Holocaust survivors lived, filed an injunction against the National Socialist Party (henceforth, the NSPA) and passed several ordinances to prevent any request to hold a neo-Nazi white power demonstration. The ordinances banned the distribution of hate speech material but reassured the organisers that the demonstrations could be held, provided they paid a substantial public safety insurance bond. One of the ordinances specifically stated that people could not wear Nazi uniforms or display swastikas during demonstrations. The NSPA organiser used both the injunction and the ordinances as an opportunity to claim infringement upon the First Amendment rights of the NSPA marchers. The Supreme Court ruled that the display of swastikas is a symbolic form of freedom of expression entitled to First Amendment protections and determined that the display of swastikas by itself did not constitute an imminent threat. As a result of the litigation, Skokie's ordinance was declared unconstitutional for its infringement upon First Amendment rights. The Supreme Court overlooked the fact that, for the survivors of the Holocaust living in Skokie, seeing the display of swastikas was interpreted as equivalent to a "physical attack", as their attorneys claimed in the legal proceedings. In other words, unlike the NSPA marchers (the ingroup), the survivors of the Holocaust (the outgroup) would perceive the public display of swastikas in the village where they lived as a salient visual ostensive stimulus (Figure 8.1). As shown below, the construction of the implicated premises is grounded in strong assumptions based on direct experiences of the horrors of the Holocaust.

Figure 8.1: The Nazi swastika.[142]

Non-linguistic explicature[143]

a. *A flag with a red background, a white disk and a Black hooked cross in the middle.*

Implicated premises

a. *Hitler created the Nazi swastika*[144] *as a symbol of the Aryan master race.*
b. *The swastika has been largely associated with Nazism, antisemitism, white supremacy and evil.*
c. *The swastika is a representative symbol of the Holocaust.*
d. *We are Holocaust survivors.*

Implicated conclusions

a. *NSPA's demonstrations in the village where Holocaust survivors live are frightening.*
b. *NSPA's demonstrations in the village where Holocaust survivors live are threatening.*

---

142 German government – RGBl. I (1935) No. 122, Public domain. https://commons.wikimedia.org/w/index.php?curid=4713270 (access 31 July 2022).

143 The question has arisen in Relevance theory literature about whether pictures can trigger explicatures (coded meanings). Forceville and Clark (2014) argue that some nonverbal behaviours can be understood as having coded meanings, which would allow for the possibility of non-linguistic explicatures.

144 See Felicity Rash (2006). *The language of violence. Adolph Hitler's Mein Kampf.* Berlin: Peter Lang.

c. *NSPA's demonstrations in the village where Holocaust survivors live are intimidating.*
d. *The public display of swastikas is like a physical attack on Holocaust survivors.*

The swastika is a salient visual ostensive stimulus that carries its presumption of relevance for the Jewish people because it produces the greatest contextual effects with the lowest cognitive processing effort. The contextual effects reinforce the assumptions the targets have about Nazis. The linguistic cues provided by the Principle of Relevance in National Socialist Party v. Skokie (1977) would support the hypothesis that, for the Jewish people, the display of Nazi swastikas implies a true threat, due to the history of terror and genocide associated with the symbol.

In Virginia v. Black (2003), the critical question was whether the physical act of cross burning was considered sufficient evidence of intimidation (*prima facie* evidence). In this case, the Supreme Court of Virginia declared unconstitutional the statute of Virginia State outlawing cross burning in public places with the intent to intimidate or place others in fear of bodily harm. The underlying issue for the Supreme Court was whether Virginia's statute violated the right to freedom of expression because of its *prima facie* provision. The Supreme Court ruled that the *prima facie* evidence provision was unconstitutional because it blurred the distinction between constitutionally banned true threats and the Klan's insignia expressing shared group identity and ideology. The Supreme Court decided that cross burning does not inevitably convey the intent to intimidate, even if the Klan had often used it as a tool of intimidation and a threat of impending violence against African Americans.

By contrast, for African Americans, cross burning on one's property or in a public place is a salient symbol of terror. In other words, cross burning is interpreted as a true threat because of the long history of violent crime perpetrated by the Klan against African Americans. Cross burning is an act of ostensive communication that automatically displays a presumption of relevance. Cross burning presupposes an intention to communicate (visual ostensive stimuli). Unlike codification-decodification communication, in which the correspondence between signs and messages is pre-established and conventional, in ostensive-inferential communication the correspondence between actions and intentions is non-conventional and must be interpreted in the context of situation. For African Americans, the most salient interpretation of cross burning (Figure 8.2) on another person's property or in a public place is intimidatory and threatening, as shown in the hypothetical deductive process displayed below.

Figure 8.2: The Ku Klux Klan burning a cross (1921). Public domain.[145]

Non-linguistic explicature
a. *Cross burning.*

Implicated premises
a. *Cross burning is long associated with the Ku Klux Klan's history of crime against African Americans.*
b. *Cross burning is associated with white supremacist views.*
c. *Many African Americans have been brutally murdered by the Ku Klux Klan.*

Implicated conclusions
a. *Cross burning is frightening.*
b. *Cross burning is intimidatory.*
c. *Cross burning is a threat of impending violence against African Americans.*

Cross burning is a salient visual ostensive stimulus that carries its presumption of relevance. The implicated premises are strong because they are based on direct experiences. The contextual effects reinforce the negative assumptions African Americans have regarding the Klan as a terrorist group. In this case, the

145 Source: The Library of Congress American Memory Collection, http://memory.loc.gov/ammem/ {{PD}} Copied from en Wikipedia.Category:Ku Klux Klan.

Principle of Relevance would support the hypothesis that, for African Americans, cross burning implies intimidation and a true threat.

The third court case that the chapter analyses is Fáber v. Hungary (2012). Here the critical question was whether the display of the Árpád-striped flag[146] (Figure 8.3) at the steps leading to the Danube embankment, where Jews had been exterminated in large numbers during the Arrow Cross Regime in Hungary, had a provocative nature and was likely to result in public disorder.

Figure 8.3: The Árpád-striped flag. Public domain.[147]

The court in Hungary found that Fáber had exceeded his right to freedom of expression, causing danger to public order. The litigation reached the European Court of Human Rights, where it was decided that the mere display of the Árpád-striped flag, a lawful flag, could not be interpreted as a true threat capable of causing public disorder. It might be noted that in this particular case, the visual ostensive stimuli used by the subject were ambiguous for various reasons:

First, although for many Hungarians the Árpád stripes represent the rich historical heritage of Hungary, the recent appropriation of the Árpád stripes both on flags and badges by Hungarian right-wing parties has raised controversy due to

146 The flag is derived from the Árpád dynasty that ruled Hungary in medieval times. In modern times, the flag is also the semi-official symbol of the Jobbik, a Christian nationalist party, founded in the beginning of the 21st c.

147 No machine-readable author provided. Svenskafan assumed (based on copyright claims). https://commons.wikimedia.org/wiki/File:Arpadflagga_hungary.png (accessed 14 August 2022).

possible fascist connotations. The reason for the controversy is that the Arrow Cross Regime, a Nazi puppet government, used a similar symbol as a component of their flag in the 1940s.[148] Apart from the ambiguity of displaying the Árpád-striped flag, the subject's behaviour was passive. He did not say or do anything other than displaying the Árpád-striped flag. Nevertheless, the place at which he was standing was interpreted by the police as an act of ostensive communication, automatically displaying a presumption of relevance: The subject was standing at the place where a large number of Jews had been massacred during the Arrow Cross Regime, at the same time members of Jobbik were expressing their disagreement with the Hungarian Socialist Party's demonstration protest against racism and hatred. The flag, together with the place where the subject was displaying it at the time of the anti-racist demonstration, led the domestic court to conclude that he was trying to provoke the participants in the anti-racist demonstration and cause public disorder. Nevertheless, neither the Hungarian Socialist Party's protesters nor the Jobbik protesters in fact reacted to the subject's ostensive stimuli. Nobody took uptake of the subject's alleged communicative intention. The ostensive stimuli selected by the subject, who had neither power nor moral authority over the audience, were not salient enough to the demonstrators and hence there was no receptiveness. The act of ostensive communication was not relevant – that is, the contextual effects were low, because they implied a high processing effort for the audience. When the ostensive stimuli the subject selects are ambiguous and equivocal, the implicated premises are weak and so are the implicated conclusions. An ambiguous ostensive act of communication also communicates a presumption of relevance that needs to be interpreted in the context of situation. One should not forget that the Principle of Relevance applies without exception and cannot be violated (Escandell-Vidal 1996: 130–133). In this case, exhibiting a passive attitude and displaying an ambiguous symbol helped Fáber have his conviction overturned by the European Court of Human Rights.

148 Although the Arrow Cross Regime flag and the Árpád-striped flag combine red and white stripes in their design, they are different in two respects: (a) the number of stripes and (b) the colour combination. Whereas the Arrow Cross Regime flag used nine stripes, starting and ending with red, the original Árpád-striped flag used eight stripes, starting with red and ending with argent. The Arrow Cross Regime flag can be seen at https://commons.wikimedia.org/wiki/File:Nyilaskereszteszaszlo.SVG (accessed 14 August 2022).

5 Conclusions

This chapter considered the interpretation of hate speech meaning from the perspective of cognitive pragmatics. At the core of the discussion was Grice's (1975) notion of implicature, because it provides the bridge from what is said to what is meant but not overtly said. The hate-advocating speaker mostly delivers information implicitly and indirectly, leaving it to the hearer to make assumptions about the speaker's intended meaning. For both linguists and legal practitioners analysing hate speech, the critical question is: How can one establish what is to be inferred?

Sperber and Wilson's (1995 [1986]) Principle of Relevance draws attention to the fact that the interpretation of the speaker's communicative intention is heuristic (non-falsifiable), which means that the hearer can only construct hypotheses. It was shown that, whether encoded in language or non-verbal, ostension plays an essential role in attracting the hearer's attention towards the most relevant interpretation of the speaker's communicative intent – that is, towards the interpretation that generates the greatest contextual effects with the lowest cognitive processing effort.

The present chapter demonstrated how malicious speakers can exploit passiveness, ambiguity and indirectness to convey their communicative intent under cover (see Fáber v. Hungary (2012)). In such cases, the speaker allows the hearer to choose between one communicative intention and another, leaving the hearer a part of the responsibility of the meaning (cf. Leech 1983: 23–24). Although passiveness and ambiguity are misleading, they also carry the presumption of optimal relevance: the speaker intends to perform a wrong act but at the same time elude their responsibility for the act.

To conclude, legal practitioners evaluating hate speech must try to balance the competing interests of a dominant group with those of a vulnerable group and make a fair decision about which interests should prevail over the other in the legal decision (cf. Delgado 2018). The Principle of Relevance provides linguistic resources to recover the speaker's communicative intention from their acts and, therefore, be helpful for legal reasoning in cases related with hate speech.

References

Ainsworth, Janet & Patrick Juola. 2019. Who wrote this? Modern forensic authorship analysis as a model for valid forensic science. *Washington University Law Review* 96 (5). 1161–1189.
Alba-Juez, Laura & J. Lachlan Mackenzie. 2016. *Pragmatics, cognition, context & culture.* Madrid: McGraw Hill & UNED.
Alston, William P. 2000. *Illocutionary acts and sentence meaning.* Ithaca: Cornell University Press.
Arndt, Horst & Richard W. Janney. 1979. Interactional and linguistic models for the analysis of speech data: An integrative approach. *Sociologia Internationalis* 17 (1–2). 3–45.
Arndt, Horst & Richard W. Janney. 1983.Towards an interactional grammar of spoken English. In John Morreal (ed.), *The Ninth LACUS Forum* 1982, 367–381. Columbia, SC: Hornbeam.
Arndt, Horst & Richard W. Janney. 1985. Improving emotive communication: Verbal, prosodic, and kinesic conflict-avoidance technique. *Per Linguam* 1 (1). 21–33.
Assimakopoulos, Stavros, Fabienne Baider & Sharon Millar. 2018. *Online hate speech in the E.U. A discourse-analytic perspective.* Switzerland: Springer Open. https://doi.org/10.1007/978-3-319-72604-5
Austin, John L. 1962. *How to do things with words.* Oxford: Oxford University Press.
Baider, Fabienne. 2020. Pragmatics lost? Overview, synthesis and proposition in defining online hate speech. *Pragmatics and Society* 11 (2). 196–218.
Baider, Fabienne. 2022. Covert hate speech, conspiracy theory and anti-Semitism: Linguistic analysis versus legal judgement. *International Journal for Semiotics of Law.* (Published online 7 April 2022). https://doi.org/10.1007/s11196-022-09882-w
Baider, Fabienne & Monika Kopytowska (eds.). 2018. Special issue on narrating hostility, challenging hostile narratives. *Lodz Papers in Pragmatics* 14 (1). 1–24.
Baker, Paul et al. 2008. A useful methodological synergy? Combining critical discourse analysis and corpus linguistics to examine discourses of refugees and asylum seekers in the UK press. *Discourse and Society* 19 (3). 273–306.
Bakhtin, Mikhail Mikhailovich. 1981. *The dialogic imagination: Four essays.* Michael Holquist (ed.). Caryl Emerson & Michael Holquist (trans.). Slavic Series 1. Austin: University of Texas Press.
Bakhtin, Mikhail Mikhailovich. 1986. The problem of speech genres. In Caryl Emerson & Michael Holquist (eds.). Vern W. McGee. (trans.), *Speech genres and other late essays,* 60–102. Austin: University of Texas Press.
Banaji, Shakuntala & Ramnath Bhat. 2022. *Social media and hate.* London: Routledge.
Bara, Bruno G., Francesca Marina Bosco & Monica Bucciarelli. 1999. Simple and complex speech acts: What makes the difference within a developmental perspective. In M. Hahn & S. C. Stoness (eds.), *Proceedings of the XXI Annual Conference of the Cognitive Science Society,* 55–60. Mahwah, NJ: Lawrence Erlbaum Associates.
Bazerman, Charles. 1994. Systems of genres and the enhancement of social intentions. In Aviva Freedman & Peter Medway (eds.), *Genre and new rhetoric,* 79–101. London: Taylor and Francis.
Bazerman, Charles. 2004. Intertextuality: How texts rely on other texts. In Charles Bazerman & Paul Prior (eds.), *What writing does and how it does it,* 309–339. New Jersey: Lawrence Erlbaum.
Becker, Matthias J. 2020. Antisemitism on the internet: An underestimated challenge requiring research-based action. *Justice* 64. 32–40.

https://doi.org/10.1515/9783110672619-009

Becker, Matthias J. 2021. *Antisemitism in reader comments: Analogies for reckoning with the past.* London: Palgrave Macmillan.
Beliaeva, Natalia. 2022. Is play on words fair play or dirty play? On ill-meaning use of morphological blending. In Natalia Knoblock (ed.), *The grammar of hate. Morphosyntactic features of hateful, aggressive, and dehumanising discourse*, 177–196. Cambridge & New York: Cambridge University Press.
Benesch, Susan. 2014. Defining and diminishing hate speech. *State of the World's Minorities and Indigenous People*, 19–25. https://minorityrights.org/wp-content/uploads/old-site-downloads/mrg-state-of-the-worlds-minorities-2014-chapter02.pdf
Berkenkotter, Carol & Thomas N. Huckin. 1995. *Genre knowledge in disciplinary communication – cognition/culture/power.* Hillsdale, NJ: Lawrence Erlbaum Associates.
Beyssade, Claire & Jean-Marie Marandin. 2006. From complex to simple speech acts: A bidimensional analysis of illocutionary forces. Proceedings of the Tenth Workshop on the Semantics and Pragmatics of Dialogue, 42–49. https://publishup.uni-potsdam.de/opus4-ubp/frontdoor/deliver/index/docId/945/file/beyssade_etal.pdf
Bhatia, Vijay K. 1993. *Analysing genre: Language use in professional settings.* London: Longman.
Bhatia, Vijay K. 1996. Methodological issues in genre analysis. *Hermes, Journal of Linguistics* 16. 39–59.
Bhatia, Vijay K. 1997. Genre missing in academic introductions. *English for Specific Purposes* 16 (3). 181–195.
Bhatia, Vijay K. 2014 [2004]. *Worlds of written discourse. A genre-based view.* London, New Delhi, New York, Sydney: Bloomsbury.
Bianchi, Robert. 2022. "Kill the invaders". Imperative verbs and their grammatical patients in Tarrant's *The Great Replacement.* In Natalia Knoblock (ed.), *The grammar of hate. Morphosyntactic features of hateful, aggressive, and dehumanising discourse*, 222–240. Cambridge & New York: Cambridge University Press.
Biber, Douglas & Susan Conrad. 2009. *Register, genre, and style.* Cambridge: Cambridge University Press.
Bick, Eckhard 2020. Annotating emoticons and emojis in a German-Danish social media corpus for hate speech research. *RASK: International Journal of Language and Communication* 52. 1–20.
Bick, Eckhard. 2021. Almans and Kanaks: Inter-cultural stereotypes in a German Twitter corpus. *RASK: International Journal of Language and Communication* 53. 101–117.
Bloor, Meriel & Thomas Bloor. 2013 [2007]. *The practice of critical discourse analysis.* 2nd ed. London & New York: Routledge.
Bitzer, Lloyd. 1968. The rhetorical situation. *Philosophy and Rhetoric* 1 (1). 1–14.
Bond, Julian, Morris Dees & Richard E. Baudouin. 2011. *Ku Klux Klan: A history of racism and violence* (6th ed.). Southern Poverty Law Center, Klanwatch Project. Montgomery, Alabama: The Southern Poverty Law Center.
Brenner, Jennifer L. 2002. True threats a more appropriate standard for analysing first amendment protection and free speech when violence is perpetrated over the internet. *North Dakota Law Review* 98 (4). 753–784.
Brindle, Andrew. 2016. *The language of hate. A corpus linguistic analysis of white supremacist language.* New York & London: Routledge.
Brown, Roger & Albert Gilman. 1960. The pronouns of power and solidarity. In Thomas A. Sebeok (ed.), *Style in language*, 252–281. Cambridge, Mass: MIT Press.

Brown, Alexander. 2017a. What is hate speech? Part 1: The myth of hate. *Law and Philosophy* 36 (4). 1–50.
Brown, Alexander. 2017b. What is hate speech? Part 2: Family resemblances. *Law and Philosophy* 36 (5). 1–53.
Brown, Alexander & Adriana Sinclair. 2020. *The politics of hate speech laws*. London & New York: Routledge.
Brown, Penelope & Stephen C. Levinson. 1987 [1978]. *Politeness. Some universals in language use*. Cambridge: Cambridge University Press.
Bruneau Emile & Nour Kteily. 2017. The enemy as animal: Symmetric dehumanization during asymmetric warfare. *PLOS ONE*. Online Open Access.
Bublitz, Wolfram. 2011. Cohesion and coherence. In Jan Zienkowski, Jan-Ola Östman & Jef Verschueren (eds.), *Discursive pragmatics*, 37–49. Amsterdam & Philadelphia: John Benjamins.
Burkhardt, Armin (ed.). 1990. *Speech acts, meaning and intentions: Critical approaches to the philosophy of John R. Searle*. Berlin & New York: De Gruyter.
Caldas-Coulthard, Carmen Rosa & Malcolm Coulthard (eds.). 1995. *Text and practices*. London: Routledge.
Chilton, Paul. 2011. Manipulation. In Jan Zienkowski, Jan-Ola Östman & Jef Verschieran (eds.), *Discursive pragmatics*, 176–189. Amsterdam, Philadelphia: John Benjamins.
Christie, Frances. 1987. Young children's writing: From spoken to written genre. *Language and Education* 1 (1). 3–13.
Cronan, John P. 2002. The next challenge for the first amendment: The framework for an internet incitement standard. *Catholic University Law Review* 51 (2). 425–466.
Culpeper, Jonathan. 1996. Towards an anatomy of impoliteness. *Journal of Pragmatics* 25. 349–367.
Culpeper, Jonathan. 2005. Impoliteness and entertainment in the television quiz show: "The Weakest Link". *Journal of Politeness Research* 1 (1). 35–72.
Culpeper, Jonathan. 2008. Reflections on impoliteness, relational work and power. In Derek Bousfield & Miriam A. Locher (eds.), *Impoliteness in language: Studies on its interplay with power in theory and practice*, 17–44. Berlin & New York: Mouton de Gruyter.
Culpeper, Jonathan. 2011. *Impoliteness. Using language to cause offence*. Cambridge: Cambridge University Press.
Culpeper, Jonathan. 2012. (Im)politeness: Three issues. *Journal of Pragmatics* 44 (9). 1128–1133.
Culpeper, Jonathan & Marina Terkourafi. 2017. Pragmatics and (im)politeness. In Jonathan Culpeper, Jonathan Haugh, Michael Kadar & Daniel Kadar (eds.), *Palgrave handbook of (im)politeness*, 11–39. Basingstoke: Palgrave.
Cruse, Alan D. 2004. *Meaning in language: An introduction to semantics and pragmatics*. Oxford: Oxford University Press.
Darquennes, Jeroen. 2015. Language conflict research: A state of the art. *International Journal of the Sociology of Language* 235. 7–32.
Davis, Steven. 1980. Perlocutions. In John R. Searle, Ferenc Kiefer & Manfred Bierwisch 1980. *Speech act theory and pragmatics*, 37–55. Dordrecht, Holland & Boston: D. Reidel Publishing Company.
De Beaugrande, Robert. 2011. Text linguistics. In Jan Zienkowski, Jan-Ola Östman & Jef Verschieran (eds.), *Discursive pragmatics*, 119–134. Amsterdam, Philadelphia: John Benjamins.

Delgado, Richard. 1982. Words that wound: A tort action for racial insults, epithets, and name-calling. *Harvard Civil Rights-Civil Liberties Law Review* 17 (1). 133–182.
Delgado, Richard & Jean Stefancic. 2018 [2004]. *Understanding words that wound.* New York and London: Routledge.
Delgado, Richard & Jean Stefancic. 2018. *Must we defend Nazis? Why the first amendment should not protect hate speech and white supremacist.* New York: New York University Press.
Dewaele, Jean-Marc. 2015. Culture and emotional language. In Farzad Sharifian (ed.), *The Routledge handbook of language and culture*, 357–370. Oxford: Routledge.
Dixon, Thomas. 1905. *The clansman.* New York: The American News Company.
Dolz Lago, María Jesús. 2015. *Los delitos de odio en el código penal tras la modificación operada por LO 1/2015.* Breve referencia a su relación con el delito del Art. 173 CP. (accessed 12 May 2022).
Dos Santos Allen, Karla. 2017. Racist speech as a linguistic discriminatory practice in Brazil: Between the speech act's reference and effects. In Daniel Silva (ed.), *Language and violence. Pragmatic perspectives*, 125–140. Amsterdam & Philadelphia: John Benjamins Publishing Company.
Dudley-Evans, Tony. 1986. Genre analysis: An investigation of the introduction and discussion sections of MSc dissertations. *Talking about text* 14. 128.
Duranti, Alessandro. 2004. Agency in language. In Alessandro Duranti (ed.), *A comparison to linguistic anthropology*, 451–473. Malden, USA: Blackwell Publishing.
Dwoskin, Elizabeth, Nitasha Tiku & Craig Timberg. 2021. Facebook's race-blind practices around hate speech came at the expense of black users, new documents show. *The Washington Post*, 21 November 2021 (accessed 31 January 2022).
Escandell-Vidal, M. Victoria. 1996. *Introducción a la pragmática.* Barcelona: Ariel.
Fairclough, Norman. 1992. *Discourse and social change.* Cambridge & Malden: Polity Press.
Fairclough, Norman. 2005. Critical discourse analysis in transdisciplinary research. In Ruth Wodak & Paul Chilton (eds.), *A new agenda in (critical) discourse analysis*, 53–70. Amsterdam & Philadelphia: John Benjamins.
Fairclough, Norman. 2010 [1995]. *Critical discourse analysis. The critical study of language.* 2nd ed. London & New York: Routledge.
Flores Ohlson, Linda. (2022). The power of a pronoun. In Natalia Knoblock (ed.), *The grammar of hate. Morphosyntactic features of hateful, aggressive, and dehumanising discourse*, 161–176. Cambridge & New York: Cambridge University Press.
Foolen, Ad. 2016. Word valence and its effects. In Ulrike M. Lüdtke (ed.), *Emotion in language*, 241–256. Amsterdam: John Benjamins.
Forceville, Charles & Billy Clark. 2014. Can pictures have explicatures? *Linguagem em (Dis)curso* 14 (3). 451–472.
Fowler, Roger, Bob Hodge, Gunther Kress & Tony Trew. 1979. *Language and control.* London: Routledge & Kegan Paul.
Fraser, Bruce. 1990. Perspectives on politeness. *Journal of Pragmatics* 14, 219–236.
Fry, Gladys-Marie. 1977. *Night riders in black folk history.* Tennessee: University of Tennessee Press.
Geyer, Klaus, Eckhard Bick & Andrea Kleene. 2022. "I am not a racist but . . .": A corpus-based analysis of xenophobic hate speech constructions in Danish and German media discourse. In Natalia Knoblock (ed.), *The grammar of hate. Morphosyntactic features of*

hateful, aggressive, and dehumanising discourse, 241–261. Cambridge & New York: Cambridge University Press.

Giltrow, Janet. 2013. Genre and computer-mediated communication. In Susan C. Herring, Dieter Stein & Tuija Virtanen (eds.), *Pragmatics of computer-mediated communication*, 717–737. Berlin: De Gruyter.

Giltrow, Janet. 2017. Bridge to genre: Spanning technological change. In Carolyn R. Miller & Ashley R. Kelly (eds.), *Emerging genres in new media environments*, 39–61. London: Palgrave Macmillan, Cham.

Giltrow, Janet & Dieter Stein (eds.). 2017. *The pragmatic turn in law. Inference and interpretation in legal discourse*. Berlin: De Gruyter Mouton.

Goffman, Ervin. 1967. *Interaction ritual: Essays on face-to-face interaction*. Chicago: Aldine.

Grice, H. Paul. 1975. Logic and conversation. In Peter Cole & Jerry L. Morgan (eds.), *Syntax and Semantics* 3, Speech Acts, 41–58. New York: Academic Press.

Grice, H. Paul. 1989. *Studies in the way of words*. Cambridge: Harvard University Press.

Griffith, David. W. 1915. *The birth of a nation*. Film directed by David W. Griffith. Released 8 February 1915. United States.

Gu Yueguo. 1990. Politeness phenomena in modern Chinese. *Journal of Pragmatics* 3. 237–257.

Gu, Yueguo. 1997. Five ways of handling a bedpan: A tripartite approach to workplace discourse. *Text* 17. 457–475.

Guillén-Nieto, Victoria. 2008. Reconciling language with culture and cognition in politeness theory. In Teresa Gibert Maceda & Laura Alba Juez (eds.), *Estudios de Filología Inglesa. Homenaje a la Dra. Asunción Alba Pelayo*, 199–217. Madrid: Universidad Nacional de Educación a Distancia.

Guillén-Nieto, Victoria, 2020. Defamation as a language crime. A sociopragmatic approach to defamation cases in the high courts of justice of Spain. *International Journal of Language and Law* (JLL) 9. 1–22.

Guillén-Nieto, Victoria. 2021. "What else can you do to pass . . . ?" A pragmatics-based approach to *quid pro quo* sexual harassment. In Janet Giltrow, Frances Olsen & Donato Mancini (eds.), *Legal meanings. The making and use of meaning in legal reasoning. Foundations in language and law series* 1. 31–55. Berlin: De Gruyter Mouton.

Guillén-Nieto, Victoria. 2022. Language as evidence in workplace harassment. *Corela* HS-36. Online.

Guillén-Nieto, Victoria & Dieter Stein. 2022. Introduction: Theory and practice in forensic linguistics. In Victoria Guillén-Nieto & Dieter Stein, *Language as evidence: Doing forensic linguistics*, 1–33. Switzerland: Palgrave Macmillan, Springer.

Guillén-Nieto, Victoria & Dieter Stein (eds.). 2022. *Language as evidence: Doing forensic linguistics*. Switzerland: Palgrave Macmillan, Springer.

Guzmán, Sebastián G. 2013. Reasons and the acceptance of authoritative speech: An empirically grounded synthesis of Habermas and Bourdieu. *Sociological Theory*, 31 (3). 267–289.

Haidt, Jonathan. (2003). The moral emotions. In Richard J. Davidson, Klaus R. Sherer & H. Hill Goldsmith (eds.), *Handbook of affective sciences*, 852–870. Oxford: Oxford University Press.

Hall, Philippa. 2019. Disability hate speech: Interrogating the online/offline distinction. In Karen Lumsden & Emily Harmer (eds.), *Online othering. Exploring digital violence and discrimination on the web*, 309–330. Switzerland: Palgrave Macmillan.

Halliday, M.A.K. 1978. *Language as social semiotic: The social interpretation of language and meaning*. London: Edward Arnold.
Halliday, M.A.K. & Ruqaiya Hasan. 1985. *Language, context, and text: Aspects of language in a social semiotic perspective*. Victoria: Deakin University Press.
Hancher, Michael. 1980. Speech acts and the law. In Roger W. Shuy & Anna Schnukal (eds.), *Language use and the uses of language*, 245–256. Washington, D. C.: Georgetown University Press.
Hareli, Shlomo & Brian Parkinson. 2008. What's social about social emotions? *Journal for the Theory of Social Behaviour* 38 (2). 131–156.
Hart, Christopher. 2007. Critical discourse analysis and metaphor: Toward a theoretical framework. *Critical Discourse Studies*, 1–22.
Herring, Susan C., Lois Ann Scheidt, Sabrina Bonus & Elija Wright. 2005. Weblogs as a bridging genre. *Information Technology & People* 18 (2). 142–171.
Herring, Susan C., Dieter Stein & Tuija Virtanen (eds.). 2013. *Pragmatics of computer-mediated communication*, 717–737. Berlin: De Gruyter.
Hickey, Leo & Miranda Stewart (eds.). 2005. *Politeness in Europe*. Clevedon, Buffalo & Toronto: Multilingual Matters Ltd.
Hofstede, Geert. 2003 [1991]. *Cultures and organizations*. London: Profile Books Ltd.
Holtgraves, Thomas. 2005. The production and perception of implicit performatives. *Journal of Pragmatics* 37. 2024–2043.
Hom, Christopher. 2008. The semantics of racial epithets. *The Journal of Philosophy* 105 (8). 416–440.
Horn, Laurence R. 1991. Given as new: When redundant affirmation isn't. *Journal of Pragmatics* 15. 305–328.
Jespersen, Otto. 1965. *Modern English grammar*. Part IV. London: Allen &Unwin.
Kaul de Marlangeon, Silvia. 1993. La fuerza de cortesía-descortesía y sus estrategias en el discurso tanguero de la década del '20. *Revista de la Sociedad Argentina de Lingüística* 3. 7–38.
Kaul de Marlangeon, Silvia. 2005. Descortesía de fustigación por afiliación exacerbada o refractariedad. In Diana Bravo (ed.), *Estudios de la (des)cortesía en español: Categorías conceptuales y aplicaciones a corpora orales y escri*tos, 299–318. Buenos Aires: Dunken.
Kaul de Marlangeon, Silvia. 2008. Impoliteness in institutional and non-institutional contexts. *Pragmatics* 18 (4). 729–749.
Kaul de Marlangeon, Silvia. 2014. Delimitación de unidades extralingüísticas de análisis del discurso de (des)cortesía. *Signo y Seña: Revista del Instituto de Lingüística* 26. 7–22.
Kaul de Marlangeon, Silvia & Laura Alba-Juez. 2012. A typology of verbal impoliteness behaviour for the English and Spanish cultures. *Revista Española de Lingüística Aplicada* 25. 69–92.
Kecskes, Istvan. 2014. *Intercultural pragmatics*. Oxford & New York: Oxford University Press.
Kendall, Gavin. 2007. What is critical discourse analysis? Ruth Wodak in Conversation with Gavin Kendall [38 paragraphs]. *Forum Qualitative Sozialforschung / Forum: Qualitative Social Research* 8 (2). Art. 29. http://nbn-resolving.de/urn:nbn:de:0114-fqs0702297. 2007 FQS http://www.qualitative-research.net/fqs/
Kienpointner, Manfred. 2011. Figures of speech. In Jan Zienkowski, Jan-Ola Östman & Jef Verschueren (eds.), *Discursive pragmatics*, 102–118. Amsterdam, Philadelphia: John Benjamins.

Knoblock, Natalia (ed.). 2022. *The grammar of hate. Morphosyntactic features of hateful, aggressive, and dehumanising discourse*. Cambridge & New York: Cambridge University Press.

Komasara, Tiffany. 2002. Planting the seeds of hatred: Why imminence should no longer be required to impose liability on internet communications. *Capital University Law Review* 29 (3). 835–856.

Kristeva, Julia. 1980. The bounded text. In Julia Kristeva, Leon S. Roudiez (ed.), Thomas Gora, Alice Jardine & Leon S. Roudiez (trans.), *Desire in language: A semiotic approach to literature and art*, 36–63. New York: Columbia University Press.

Lakoff, Robin. 1973. The logic of politeness: Or, minding your p's and q's. In Claudia Corum, T. Cedric Smith-Stark & Ann Weiser (eds.), *Papers from the Ninth Regional Meeting of the Chicago Linguistic Society*, 292–305. Chicago: Chicago Linguistic Society.

Landa Gorostiza, Jon-Mirena. 2018. *Los delitos de odio*. Artículos 510 y 22. 4º CP 1995. Valencia: Tirant lo Blanch.

Larsen, Justine S. I. 2018. Rise of the KKK: Political rhetoric of the 1920s Ku Klux Klan in the West. *Undergraduate Honors Capstone Projects*. 430. https://digitalcommons.usu.edu/honors/430

Leech, Geoffrey N. 1983. *Principles of pragmatics*. London: Longman.

Leech, Geoffrey N. 2014. *The pragmatics of politeness*. Oxford: Oxford University Press.

Leezenberg, Michiel. 2017. Free speech, hate speech, and hate beards: Language ideologies of Dutch populism. In Daniel Silva (ed.), *Language and violence. Pragmatic perspectives*, 141–168. Amsterdam & Philadelphia: John Benjamins Publishing Company.

Leuchtenburg, William E. 1973. A Klansman joins the Court: The appointment of Hugo L. Black. *The University of Chicago Law Review* 41 (1). 1–31.

Levinson, Stephen. 1983. *Pragmatics*. Cambridge: Cambridge University Press.

Levinson, Stephen. 2000. *Presumptive meanings: The theory of generalized conversational implicature*. Cambridge, MA & London: MIT Press.

Lind, Miriam & Damaris Nübling. 2022. The neutering neuter. The discursive use of German grammatical gender in dehumanisation. In Natalia Knoblock (ed.), *The grammar of hate. Morphosyntactic features of hateful, aggressive, and dehumanising discourse*, 118–139. Cambridge & New York: Cambridge University Press.

Locher, Miriam A. & Derek Bousfield. 2008. Introduction: Impoliteness and power in language. In Derek Bousfield & Miriam A. Locher (eds.), *Impoliteness in language: Sudies on its interplay with power in theory and practice*, 1–13. Berlin & New York: Mouton de Gruyter.

Lounsbury, Fred C. 1907. "He's a Yellow Peril Chink . . .," *Chinese Immigration in the Late 19th Century*. (accessed 28 July 2022).

Lumsden, Karen & Emily Harmer (eds.). 2019. *Online othering. Exploring digital violence and discrimination on the web*. Switzerland: Palgrave Macmillan.

Lutgen-Sanvik, Pamela & Sarah J. Tracy. 2012. Answering five key questions about workplace bullying: how communication scholarship provides thought leadership for transforming abuse at work 26 (3). 3–47.

Machin, David & Andrea Mayr. 2012. *How to do critical discourse analysis*. London: Sage.

Malle, Bertram F. & Joshua Knobe. 1997. The folk concept of intentionality. *Journal of Experimental Social Psychology* 33 (2). 101–102.

Martin, James Robert. 1993. A contextual theory of language. In Mary Kalantzis & Bill Cope (eds.), *The powers of literacy: A genre approach to teaching writing*, 116–136. Pittsburgh: University of Pittsburgh Press.

Martin, James Robert, Christie Frances & Joan Rothery. 1987. Social processes in education: A reply to Sawyer and Watson (and others). In Ian Reid (ed.), *The place of genre in learning: Current debates*, 46–57. Geelong & Australia: Deakin University Press.

Marwick, Alice & Ross Miller. 2014. Online harassment, defamation, and hateful speech: A primer of the legal landscape. Fordham Center on Law and Information Policy (*Report* June 10, 2014), 1–73.

Massey, Calvin. 1992. Hate speech, cultural diversity and the foundational paradigms of free expression. *UCLA Law Review* 40 (1). 103–198.

Mattiello, Elisa. 2022. Language aggression in English slang: The case of the -o suffix. In Natalia Knoblock (ed.), *The grammar of hate. Morphosyntactic features of hateful, aggressive, and dehumanising discourse*, 34–58. Cambridge & New York: Cambridge University Press.

Matsuda, Mari J. 1989. Public response to racist speech: Considering the victim's story. *Michigan Law Review* 87. 2320–2381.

Matsuda, Mari J., Charles Lawrence III, Richard Delgado, Kimberle Williams Crenshaw. 1993. *Words that wound. Critical race theory, assaultive speech, and the first amendment.* New York and London: Routledge.

McAndrew, Tara. 2017. The history of the KKK in American politics. *JSTOR Daily*, January 25th, 2017. (accessed 04 October 2021).

Metzl, Jonathan M. 2019. What guns mean: The symbolic lives of firearms. *Palgrave Communications* 5(35). 1–5.

Mey, Jacob L. & Mary Talbot. 1988. Computation and the Soul. *Journal of Pragmatics*. 12 (5–6). 743–789.

Millar, Sharon. 2019. Hate speech: Conceptualisations, interpretations and reactions. In Matthew Evans, Lesley Jeffries & Jim O'Driscoll (eds.), *The Routledge handbook of language in conflict*, 145–163. Oxon & New York: Routledge.

Miller, Carolyn R. 1984. Genre as social action. *Quarterly Journal of Speech* 70 (2). 151–67.

Mirón, Louis F. & Jonathan Xavier Inda. 2000. Race as a kind of speech act. *Cultural Studies: A Research Annual* 5. 85–107.

Monnier, Angeliki, Axel Boursier & Annabelle Seoane. 2022. *Cyberhate in the context of migrations*. Switzerland: Palgrave Macmillan, Springer.

Moran, Mayo. 1994. Talking about hate speech: Rhetorical analysis of American and Canadian approaches to the regulation of hate speech. *Wisconsin Law Review* 6. 1425–1514.

Moscovici, Serge. 1982. The coming era of representations. In Jean-Paul Codol & Jacques-Philippe-Leyens (eds.), *Cognitive analysis of social behaviour*, 115–150. The Hague: Martinues Nijhoff Publishers.

Motsch, Wolgang. 1980. Situational context and illocutionary force. In John R. Searle, Ferenc Kiefer & Manfred Bierwisch (eds.), *Speech act theory and pragmatics*, 155–168. Dordrecht, Holland & Boston: D. Reidel.

Muschalik, Julia. 2018. *Threatening in English: A mixed method approach*. Amsterdam & Philadelphia: John Benjamins.

Musolff, Andreas. 2012. The study of metaphor as part of critical discourse analysis. *Critical Discourse Studies* 9 (3). 301–310.

Nelde, Peter H. 1987. Language contact means language conflict. *Journal of Multilingual and Multicultural Development* 8 (1–2). 33–42.

Nix, Naomi & Lauren Etter. 2021. *The Print*, 25 October 2021 (accessed 31 January 2022).

O'Connor, Catherine & Sarah Michaels. 2007. When is dialogue "dialogic"? *Human Development* 50 (5). 275–285.

Östman, Jan-Ola & Tuija Virtanen. 2011. Text linguistics. In Jan Zienkowski, Jan-Ola Östman & Jef Verschueren (eds.), *Discursive pragmatics*, 266–285. Amsterdam & Philadelphia: John Benjamins.

Parekh, Bhikhu. 2006. Hate speech? Is there a case for banning? *Public Policy Research* 12 (4). 2013–2023.

Peterson, David. 2022. Homophobic space-times: Lexico-grammatical and discourse-semantic aspects of the softscapes of hate. In Natalia Knoblock (ed.), *The grammar of hate. Morphosyntactic features of hateful, aggressive, and dehumanising discourse*, 262–287. Cambridge & New York: Cambridge University Press.

Philips, Susan U. 2015. Language ideologies. In Deborah Tannen, Deborah Schiffrin & Heidi Hamilton (eds.), *Handbook of discourse analysis*, 557–575. UK: Wiley Blackwell.

Philips, Susan U. 2004. Language and social inequality. In Alessandro Duranti (ed.), *A companion to linguistic anthropology*, 474–495. Malden, MA: Blackwell.

Pitruzzella, Giovanni & Pollicino, Oreste. 2020. *Disinformation and hate speech. A European constitutional perspective*. Milan: Bocconi University Press.

Quirk, Randolph, Sidney Greenbaum, Geoffrey N. Leech & Jan Svartvik. 1985. *A comprehensive grammar of the English language*. London: Longman.

Rash, Felicity. 2006. *The language of violence. Adolph Hitler's Mein Kampf*. Berlin: Peter Lang.

Recanati, François. 1980. Some remarks on explicit performatives, indirect speech acts, locutionary meaning and truth value. In John R. Searle, Ferenc Kiefer & Manfred Bierwisch (eds.), *Speech act theory and pragmatics*, 205–220. Dordrecht, Holland & Boston: D. Reidel.

Reeck, Crystal, Daniel Ames & Kevin N. Ochsner. 2016. The social regulation of emotion: An integrative, cross-disciplinary model. *Trends in Cognitive Sciences* 20 (1). 47–63.

Reisigl, Martin & Ruth Wodak. 2001. *Discourse and discrimination. Rhetoric of discrimination and antisemitism*. London & New York: Routledge.

Ruzaite, Jurate. 2018. In search of hate speech in Lithuanian public discourse: A corpus-assisted analysis of online comments. *Lodz Papers in Pragmatics* 14 (1). 93–116.

Sadock, Jerrold. 1978. On testing for conversational implicature. In Peter Cole (ed.), *Syntax and semantics (Pragmatics)* 9, 281–298. New York: Academic Press.

Sadock, Jerrold. 2007. Speech acts. In Lawrence R. Horn & Gregory Ward (eds.), *The handbook of pragmatics*, 53–73. Oxford: Blackwell.

Sarangi, Skrikant. 2011. Public discourse. In Jan Zienkowski, Jan-Ola Östman & Jef Verschuerekn (eds.), *Discursive pragmatics*, 248–265. Amsterdam & Philadelphia: John Benjamins.

Saxe, John Godfrey. 1879. *The poems of John Godfrey Saxe*. Boston: James R. Osgood and Company.

Searle, John R. 1969. *Speech acts*. Cambridge: Cambridge University Press.

Searle, John R. 1975. Indirect speech acts. In Peter Cole & Jerry L. Morgan (eds.), *Syntax and semantics: Speech acts* 3. 59–82. New York: Academic Press.

Searle, John R. 1979. *Expression and meaning. Studies in the theory of speech acts*. Cambridge. Cambridge University Press.

Searle, John R. 1980. The background of meaning. In John R. Searle, Ferenc Kiefer & Manfred Bierwisch (eds.), *Speech act theory and pragmatics*, 221–232. Dordrecht: Holland, Boston: U.S.A.: D. Reidel Publishing Company.

Searle, John R., Ferenc Kiefer & Manfred Bierwisch. 1980. *Speech act theory and pragmatics.* Dordrecht, Holland & Boston: D. Reidel.
Searle, John, R. & Daniel Vanderveken. 1985. *Foundations of illocutionary logic.* Cambridge: Cambridge University Press.
Sellars, Andrew. 2016. *Defining hate speech.* Boston Univ. School of Law, Public Law Research Paper No. 16–48. Available at SSRN: https://ssrn.com/abstract=2882244
Shaver, Phillip, Judith Schwartz, Donald Kirson & Cary O'Connor. 1987. Emotion knowledge: Further exploration of a prototype approach. *Journal of Personality and Social Psychology* 52 (6). 1061–1086.
Shuy, Roger W. 2010. *The language of defamation cases.* Oxford & New York: Oxford University Press.
Silva, Daniel. 2017. The circulation of violence in discourse. In Daniel Silva (ed.), *Language and violence. Pragmatic perspectives*, 107–124. Amsterdam & Philadelphia: John Benjamins.
Silva, Daniel (ed.). 2017. *Language and violence. Pragmatic perspectives.* Amsterdam & Philadelphia: John Benjamins.
Solin, Anna. 2011. Genre. In Jan Zienkowski, Jan-Ola Östman & Jef Verschieran (eds.), *Discursive pragmatics*, 119–134. Amsterdam & Philadelphia: John Benjamins.
Spencer-Oatey, Helen. 2000. Rapport management: A framework for analysis. In Helen Spencer-Oatey (ed.), *Culturally speaking: Managing rapport through talk across cultures*, 11–46. London & New York: Continuum.
Spencer-Oatey, Helen. 2008 [2000]. *Culturally speaking: culture, communication and politeness theory.* London & New York: Continuum.
Sperber, Dan & Deirdre Wilson. 1995 [1986]. *Relevance. Communication & cognition.* 2nd ed. Oxford UK & Cambridge USA: Blackwell.
Stein, Dieter. 2022. Mobbing as genre and cause for legal action? Linguistic prolegomena for a legal issue. *Corela* HS-36. Online.
Stollznow, Karen. 2017. *The Language of discrimination.* Munich: LINCOM GmbH.
Strani, Katerina & Anna Szczepaniak-Kozak. 2018. Strategies of othering through discursive practices: Examples from the UK and Poland. *Lodz Papers in Pragmatics* 14 (1). 163–179.
Swales, John M. 1990. *Genre analysis: English in academic and research settings.* Cambridge: Cambridge University Press.
Tannen, Deborah, Heidi E. Hamilton & Deborah Schiffrin (eds.). 2015. *The handbook of discourse analysis.* Oxford: Blackwell.
Tarasova, Elizaveta & José A. Sánchez Fajardo. 2022. Adj + ie/ly nominalization in contemporary English: From diminution to pejoration. In Natalia Knoblock (ed.), *The grammar of hate. Morphosyntactic features of hateful, aggressive, and dehumanising discourse*, 59–81. Cambridge & New York: Cambridge University Press.
Technau, Björn. 2018. Going beyond hate speech: The pragmatics of ethnic slur terms. *Lodz Papers in Pragmatics* 14 (1). 25–43.
Terkourafi, Marina. 1999. Frames for politeness: A case study. *Pragmatics* 9 (1). 97–117.
Terkourafi, Marina. 2016. The linguistics of politeness and social relations. In Keith Allan (ed.), *The Routledge handbook of linguistics*, 221–235.Oxford: Routledge.
Thál, Jonáš & Irene Elmerot. (2022). Unseen gender: Misgendering of transgender individuals in Czech. In Natalia Knoblock (ed.), *The grammar of hate. Morphosyntactic features of hateful, aggressive, and dehumanising discourse*, 97–117. Cambridge & New York: Cambridge University Press.

Thomas, Jenny. 1995. *Meaning in interaction: An introduction to pragmatics*. London: Longman.
Thomson, Geoff & Laura Alba-Juez (eds.). 2014. *Evaluation in context*. Amsterdam: John Benjamins.
Tiersma, Peter M. 1987. The language of defamation. *Texas Law Review* 66 (2). 303–350.
Toolan, Michael. 1997. What is critical discourse analysis and why are people saying such terrible things about it? *Language and Literature* 6 (2). 83–103.
Törnberg, Anton & Petter Törnberg. 2016. Muslims in social media discourse: Combining topic modelling and critical discourse analysis. *Discourse, Context and Media* 13. 132–142.
Tsesis, Alexander. 2002. *Destructive messages. How hate speech paves the way for harmful social movements*. New York & London: New York University Press.
Tsesis, Alexander. 2009. Dignity and speech: The regulation of hate speech in a democracy. *Law Review*, 497–532.
Udupa, Sahana, Iginio Gagliardone & Peter Hervik (eds.). 2021. *Digital hate. The global conjuncture of extreme speech*. Indiana: Indiana University Press.
van Dijk, Teun A. 1981. Episodes as units of discourse analysis. In Deborah Tannen (ed.), *Analysing discourse: Text and talk*, 177–195. Georgetown: Georgetown University Press.
van Dijk, Teun A. 1992. Discourse and the denial of racism. *Discourse & Society* 3 (1). 87–118.
van Dijk, Teun A. 1995a. Discourse analysis and ideology analysis. In Christina Schäffne & Anita L. Wenden (eds.), *Language and peace*, 17–33. London: Routledge.
van Dijk, Teun A. 1995b. Discourse semantics and ideology. *Discourse & Society* 6 (2). 243–289.
van Dijk, Teun A. 1995c. Aims of critical discourse analysis. *Japanese Discourse* 1. 17–27.
van Dijk, Teun A. 1996. Discourse, power and access. In Carmen Rosa Caldas-Coulthard & Malcolm Coulthard (eds.), *Text and practices*, 84–104. London: Routledge.
van Dijk, Teun A. 1997. Context models and text processing. In Max Stamenow (ed.), *Language structure, discourse and the access to consciousness*, 189–226. Amsterdam: Benjamins.
van Dijk, Teun A. 2005. Contextual knowledge management in discourse production. A CDA perspective. In Ruth Wodak & Paul Chilton (eds.), *A new agenda in (critical) discourse analysis*, 80–100. Amsterdam & Philadelphia: John Benjamins.
van Dijk, Teun A. 2006a. Discourse and manipulation. *Discourse & Society* 17 (2). 359–383. doi:
van Dijk, Teun A. 2006b. Discourse, context and cognition. *Discourse Studies* 8 (1). 159–177.
van Dijk, Teun A. 2011. Discourse and ideology. In Teun A. van Dijk (ed.), *Discourse studies: A multidisciplinary introduction*, 379–407. Thousand Oaks, California, US: Sage Publications Limited.
van Dijk, Teun A. 2015. Critical discourse analysis. In Deborah Tannen, Deborah Schiffrin & Heidi E. Hamilton (eds.), *Handbook of discourse analysis*, 466–485. Oxford: Blackwell.
van Leeuwen, Theo. 2005. Three models of interdisciplinarity. In Ruth Wodak & Paul Chilton (eds.), *A new agenda in (critical) discourse analysis*, 3–18. Amsterdam, Philadelphia: John Benjamins.
van Leeuwen, Theo. 2015. Multimodality. In Deborah Tannen, Deborah Schiffrin & Heidi E. Hamilton (eds.), *Handbook of discourse analysis*, 447–465. Oxford: Blackwell.
Waldron, Jeremy. 2012. *The harm in hate speech*. Cambridge, Massachusetts & London, England: Harvard University Press.

Ward, Kenneth. 1997. Free speech and the development of liberal virtues: An examination of the controversies involving flag-burning and hate speech. *University of Miami Law Review* 52 (3). 733–792.
Watts, Richard J. 2003. *Politeness*. Cambridge: Cambridge University Press.
Wilson, Deirdre & Dan Sperber. 2007. Relevance theory. In Laurence R. Horn & Gregory Ward (eds.), *The handbook of pragmatics*, 607–632. London: Blackwell.
Winter, Aaron. 2019. Online hate: From the far-right to the "alt-right" and from the margins to the mainstream. In Karen Lumsden & Emily Harmer (eds.), *Online othering. Exploring digital violence and discrimination on the web*, 39–63. Switzerland: Palgrave Macmillan.
Wittgenstein, Ludwig. 2009 [1953]. *Philosophical investigations* (4th ed.). P. M. S. Hacker & Joachim Schulte (eds.). G. E. M. Anscombe, P. M. S. Hacker & Joachim Schulte (trans.). Oxford: Blackwell.
Wodak, Ruth. 2001. The discourse-historical approach. In Ruth Wodak & Michael Meyer (eds.)., *Methods of critical discourse analysis*, 63–94. London: Sage.
Wodak, Ruth. 2011. Critical linguistics and critical discourse analysis. In Jan Zienkowski, Jan-Ola Östman & Jef Verschueren (eds.), *Discursive pragmatics*, 50–70. Amsterdam & Philadelphia: John Benjamins.
Wodak, Ruth. 2021 [2015]. *The politics of fear. The shameless normalisation of far-right discourse* (2nd ed.). London: Sage.
Wodak, Ruth & Paul Chilton (eds.). 2005. *A new agenda in (critical) discourse analysis*. Amsterdam & Philadelphia: John Benjamins.
Wodak, Ruth & Martin Reisigl. 2015. Discourse and racism. In Deborah Tannen, Deborah Schiffrin & Heidi E. Hamilton (eds.), *Handbook of discourse analysis*, 576–596. Oxford: Blackwell.
Wodak, Ruth & Michael Meyer. 2016 [2001]. Critical discourse analysis: History, agenda, theory, and methodology. In Ruth Wodak & Michael Meyer (eds.), *Methods for critical discourse analysis* (3rd ed.), 1–33. London: Sage.
Wodak, Ruth & Michael Meyer (eds.). 2016 [2001]. *Methods of critical discourse analysis* (3rd ed.). London: Sage.
Zienkowski, Jan, Jan-Ola Östman & Jef Verschueren (eds.). 2011. *Discursive pragmatics*. Amsterdam & Philadelphia: John Benjamins.

Legal references

Bill of Rights. 1688. Online. (accessed 28 August 2020).
Constitution of the United States. First Amendment. Online. (accessed 27 May 2020).
Council Framework Decision 2008/913/JHA. 2008. *Official Journal* (OJ L 328 of 6 December 2008). *Council of Europe Committee of Ministers*. Recommendation No. R (97) 20 of the Committee of Ministers to the Member States on Hate Speech. Online. (accessed 27 May 2020).
Council of Europe. No Hate Speech Movement. Online. (accessed 27 May 2020).
Criminal Code. 2016. Online. (accessed 17 August 2022).
Criminal Code. 1998. Online. (accessed 29 May 2020).
Criminal Code. R.S.C. 1985. Online. (accessed 27 May 2020).

European Commission against Racism and Intolerance (ECRI). General Policy Recommendation No. 7 (revised) on national legislation to combat racism and rational discrimination. 2002. (revised on 7 December 2017). Online. (accessed 31 May 2020).
European Commission against Racism and Intolerance (ECRI). General Policy Recommendation No. 15 on combating hate speech. Online. (accessed 10 August 2020).
European Convention on Human Rights. Convention for the Protection of Human Rights and Fundamental Freedoms. 1950. Online. (accessed 28 August 2020).
European Court of Human Rights. Jersild v. Denmark. 1994. Application nº 15890/89. Judgment. Strasbourg.
European Court of Human Rights. A. v. The United Kingdom. 2003. Application nº 35373/97. Judgment. Strasbourg.
European Court of Human Rights. Fáber v. Hungary. 2012. Application nº 40721/08. Judgment. Strasbourg.
European Court of Human Rights. Case of ES v. Austria. 2019. Application nº 38450/12. Judgment. Strasbourg.
Fighting words, hostile audiences and true threats: Overview. Online. (accessed 30 July 2020).
Hate Crime Statistics Act, As Amended, 28 USC. § 534 § [Sec. 1.]. 2017. Online. (accessed 27 May 2020).
Hate speech – European Court of Human Rights – Council of Europe. Online. (accessed 3 June 2020).
International Convention on the Elimination of All Forms of Racial Discrimination. 1965. Online. (accessed 3 June 2020).
International Covenant on Civil and Political Rights. 1966. Online. (accessed 26 May 2020).
Loi du 29 Juillet 1881 sur la Liberté de la Presse. 1881. Version consolidée au 08 Juillet 2014. Online. (accessed 29 May 2020).
Organic Law 18/2007, of 11 July, Against Violence, Racism, Xenophobia and Intolerance in Sport. 2007. Online. (accessed 30 May 2020).
Organic Law 1/2015 of 30 March 2015, on Amendments to the Penal Code. (Organic Law 10/1995 of 23 November 1995). Online. (accessed 30 May 2020).
Public Order Act. 1986. https://www.legislation.gov.uk/ukpga/1986/64. (accessed 24 July 2020).
Rabat Plan of Action on the Prohibition of Advocacy of National, Racial or Religious Hatred that Constitutes Incitement to Discrimination, Hostility or Violence. Online. (accessed 10 August 2020).
Race Relations Act. 1968. Online. (accessed 27 May 2020).
Racial Discrimination Bill. 1975. Online. (accessed 28 May 2020).
Report of the United Nations High Commissioner for Human Rights on the expert workshops on the prohibition of incitement to national, racial or religious hatred. 2013. Online. (accessed 10 August 2020).
Spanish Constitution. 1978. Online. (accessed 30 May 2020).
Spanish Penal Code. 1995. Online. (accessed 3 June 2020).
Spanish Penal Code. 2005. English translation. With the participation of John Rason Spencer QC Professor of Law, University of Cambridge, Fellow of Selwyn College. Online. (accessed 29 May 2020).
The Racial Vilification Act. 2004. Online. (accessed 28 May 2020).
Thomson, Alec. 1940. Smith Act. *The First Amendment Encyclopedia*. Online. (accessed 17 October 2021).

Transcript of Trump Speech at Rally before US Capitol Riot. Online. (accessed 17 October 2021).

United States Constitution Law. First Amendment. Online. (accessed 27 May 2020).

Universal Declaration of Human Rights (UDHR). 1948. Online. (accessed 3 June 2020).

US Court of Appeals for the Armed Forces. United States v. Wilcox. 2008. № 05-0159. Online. (accessed 19 August 2020).

US Court of Appeals. Collin v. Smith. 1978. No. 578 F.2d 1197. Online. (accessed 3 June 2020).

US Court of Appeals. United States v. Alkhabaz. 1997. No. 104 F.3d. 1492. Online. (accessed 6 May 2021).

US Court of Appeals. Purtell v. Mason. 2008. No. 06-3176. Online. (accessed May 2022).

US Supreme Court. Schenck v. the United States. 1919. No. 249 US, 47. Online. (accessed 20 August 2020).

US Supreme Court. Chaplinsky v. New Hampshire. 1942. No. 255. Online. (accessed 20 May 2022).

US Supreme Court. Terminiello v. Chicago. (1949). No. 272. Online. (accessed 18 August 2020).

US Supreme Court. Dennis v. United States. 1951. No. 341 US 494. Online. (accessed 17 October 2021).

US Supreme Court. Brandenburg v. Ohio. 1969. No. 395 US 4444. Online. (accessed 18 August 2020).

US Supreme Court. National Socialist Party v. Skokie. 1977. No. 76-1786. Online. (accessed 18 August 2020).

US Supreme Court. Virginia v. Black. 2003. No. 538 US 343. Certiorari to the Supreme Court No. 01-1107. Online. (accessed 24 August 2020).

Index

https://doi.org/10.1515/9783110672619-010